A GUIDE TO CEDAR GLADES

and

Common Appalachian Wildflowers

Billy Plant III

For truth, my honored Tennessee friends, go and see, and learn to appreciate and to preserve such great ornaments of your native land.
                                                    - Augustin Gattinger (1901)

...to impartial science the humblest weed
Is as immortal once as the proudest flower.
                                    from "Tall Ambrosia"
                                    Henry David Thoreau

Acknowledgements

I would like to thank the many people along the way who have both directly and indirectly helped make this field guide possible. Many thanks to Drs. Kurt Blum, Thomas Hemmerly, and Jeff Walck for teaching me about the incredible variety of plants just outside my own backdoor. Much appreciation goes to Roger Shaner for putting together the fund raising video that helped secure the capital that made the physical copy of this field guide possible. Thanks to Alex Pollock and the entire staff of Pollock Printing for turning a large pdf file into a tangible volume that can be taken into the field and enjoyed for years to come. Thanks to Vicky Kremer for her love and companionship during the assembling of this work and for her amazing line drawings that bring the Flower and Leaf Morphology section to life. And to Mazy, my husky/Australian shepherd dog, who has walked with me every step of the way for the three years I have worked on this project.

And finally I want to thank my parents, Billy and Linda Plant for their guidance, love, and patience. I dedicate this book to them.

Table of Contents

Preface

Cedar glades are naturally occurring landscapes of the Central South. Geographically this includes northern Alabama and Georgia, and parts of southern Kentucky. However they reach their peak development and boast the greatest number of endemic species in the Central Basin of Tennessee. Cedar glades are characterized by open areas of thin soils and gravelly limestone surrounded by forests of oak, hickory, winged elm, blue ash, and most typically, Eastern redcedars. They have captured the attention of ecologists due to the harsh environment to which cedar glade plants have adapted. Winters are cold and wet, with soils being saturated for weeks on end. In summer the glades dry up, becoming desert like. Temperatures in open glades can be up to thirty degrees higher than in the surrounding woods. It is in the Spring, in the weeks between the harsh extremes of winter and summer, that cedar glades blossom and give rise to a colorful array of wildflowers. The highly localized common names of Nashville mustard, Nashville breadroot, Gattinger's prairie clover, and Tennessee coneflower give some indication of the limited range of many of the glades' most beautiful plants. Through spring and early summer these plants and dozens of others bloom to give a blush of color to the otherwise gray backdrop of the fissured and crumbling limestone substrate from which they arise.

The rapid development the region has experienced in the last forty years has threatened the habitat of these rare plant communities causing several species to become endangered. For example, in the 1970's one of the last naturally occurring stands of Tennessee coneflower was destroyed so a trailer park could be built on the site. Fortunately this wildflower was saved by the hard work of a handful of dedicated individuals and in 2011 Tennessee coneflower was removed from the endangered species list. The responsibility of preserving rare landscapes is one we all share. But to care about saving anything we must first understand it. I hope this field guide is a facilitator in that process.

So strap on a daypack, limber up your limbs, and let's take a walk in the glades to see what we can find...and be sure to bring a bottle of water!

Billy Plant III
February 2, 2014

1

A Year In the Tennessee Cedar Glades

A cold January wind blew across the rocky openness of the cedar glades, coupled with a fine mist, as my friend Jason and I hiked into Flat Rock Cedar Glades and Barrens State Natural Area to layout the groundwork for my thesis research. Sloshing through the mud of a wet weather creek the force of our tramping splattered mud as high as the knees on my worn out Carhartt pants. Carolina chickadees and a few cardinals gave chirping commentary as we walked about on this day when no others ventured into the desolate glades.

We arrived at the first glade and sat down our back packs to reconnoiter the site for the best location. We had come to transfer work I had already done in my mind into reality by driving ten-inch galvanized nails into the ground. I'd brought Jason, a carpenter and former biological technician with the U.S. Forest Service, to help me. With his long goatee, long hair, and wire-rimmed spectacles he looked as much like a Confederate cavalry officer or jam band musician as a botanist.

I had finalized the study design on paper a month before, and for the most part I knew which areas of the eight hundred acre nature preserve we would be going to. I would have two transects in each of the chosen glades. One transect would run through a wet zone, the other through a corresponding dry zone. The goal of the study was to identify whether plant communities and properties associated with them such as species richness and coverage were affected by soil depth and moisture availability.

To begin work I randomly chose a starting point at the bottom of a gentle slope and drove a nail into the gravelly ground as far as it would go before hitting limestone bedrock…about four inches. Then I had Jason hold the zero end of a tape measure on the nail as I walked out fifteen meters along a path where imbricated rocks and swept vegetation indicated water flowed during a rain event. At this point I drove another nail, stacking limestone flagstones around it to help the nail stay in place in the shallow soil. I tied some florescent pink flagging around each nail to make them easier to see. After this I used a compass to measure the bearing of the transect starting at the lowest point. Next we walked about fifteen yards uphill to the driest part of the glade. Once again I randomly chose a starting point to drive the first nail of the dry transect. We then set the endpoint of the dry transect along the same bearing we had just recorded for the wet transect. By doing this we ensured the transects ran parallel to one another and would cross the same general features of the glade.

The nails set, I recorded the distances, bearings, and rough GPS coordinatesof each transect into a log book. Then we trudged back to the mucky trail and on to the next site. That day we established six of the ten transects that would provide the backbone for my study. Setting the precedent for what would become a tradition, on the way back into town we celebrated our days work with a couple of beers at the "cleanest little dive bar this side of the river".

*   *   *

Cold, snowy January melted into the intermittent drizzle, flurries, and occasionally promising sunshine of February. Late in the month on a bright fifty-five degree day my friend Meadow and I tramped out to the glades to relieve cabin fever. The sun felt warm and glowing on our light deprived faces. Our dogs ran through the woods all around us as we made the long trek back to one of my more remote study sites. The transect ran through what cedar glade ecologists call a limestone glade streamside meadow. A little spring at the edge of the woods sent forth a small stream which trickled through long fissures in an otherwise solid slab of limestone. At the end of the slab a thin layer of soil and organic muck had accumulated and on this day we were happy to find a profusion of Nashville mustard (*Leavenworthia stylosa*) which had sprouted in early winter and was nearly in bloom. A few of the little white flowers with their yellow centers had already opened up revealing their sweet fragrance.

I moved a couple of flagstones onto the limestone slab for us to sit on. Meadow is a pretty hippie girl with big doe eyes and long brown hair. Having grown up on a nearby farm she was familiar with Nature but was often amazed at the wonders I would point out in her own backyard. Nothing in Nature is mundane. If you look closely enough you will see there is function to form. The wonder comes in figuring out what benefit a particular flower gains by adopting one form over another. The *Leavenworthia* is a case in point. Why are some populations white while others are yellow while other populations contain both white and yellow flowers? Before I began training as a botanist a flower had to stand on the merits of its beauty alone. But the more I learn the more I enjoy looking at the "pretty flower" and putting a name to it.

I sometimes wonder if the incessant labeling by botanists, ecologists, and others of our ilk doesn't take the wonder out of the world. After all, aren't many of the greatest artistic statements, whether they be fables or paintings, stylized mythologies created to explain reality? But I always find myself looking to identify the flowers in a painting or the historic precedent behind Shakespeare's plays. I've read the Gothic tales of the Romantics such as Hawthorne and Poe and many parables from the world's great religions and I always prefer what is real, made of stone and wood, carbon and chloroplasts. The creative imagination can be used to gain insight into reality just as much as it can be used to create fantastic worlds that could never actually exist.

I pulled my Swiss army knife out of my pocket and opened a bottle of pinot noir. I walked Meadow around the larger glade showing her my transects and telling her how they worked. She picked up the brown and orange shell of a box turtle and carried it in her hand as she wore a little sprig of the *Leavenworthia* behind her ear. Ecology is a discipline of observations and statistics but at its most basic level knowledge of any sort is only real when it is tangible. Keep the spreadsheet of crunched numbers. I'll take the muddy boots and wildflowers on a walk with a pretty girl any day.

We walked back to the slab where I built a little deadwood fire on the rock as the afternoon faded and the air grew chill. Not only were these glades at Flat Rock the site of my research, they were becoming my sanctuary, a special place I could go to escape or occasionally share with a good friend.

*   *   *

On a grey day in early March I walked into Flat Rock to begin my first round of data collection. The air was cold and the thick blanket of clouds looked as though they might release flurries or a fine mist at any moment. A rabbit jumped in the cedar at the edge of the woods and my dog ran off after it. Arriving at my first transect I pulled out the tape measure between the two nails that marked the ends. I then laid down a 1.0 x 0.5 square meter PVC plot I had built. The object was to measure soil depth and take reading of its moisture content at specified locations within the plot. For the soil moisture readings I used a probe on loan from the U.S. Geological Survey office in Nashville. After taking these reading at three locations within the plot I got down on my knees to identify the plant species found and estimate how much of the plot each species covered. Being adept at plant identification I thought this would be an easy enough task. But plants in early spring often hide the secrets to their identity in underdeveloped leaves that must unfurl and grow in response to the light of longer days before they can truly be known. For these unidentifiable sprouts I noted their location within the plot in my field book. Still other plants were easy to identify: the *Leavenworthia* which was now in full bloom, the basal rosettes of slender fleabane (*Erigeron strigosus*), and the nodding onion (*Allium cernuum*) peeping through the muck. I also found glade moss (*Pleurocheata squarrossa*) and *Nostoc* in abundance. All else was limestone and organic debris. The work was slow going. I had ten transects with five plots along each transect. For the sake of uniformity I needed to get all of them monitored within a three or four day window. After four hours the clouds opened up releasing a cold drizzle.

By Saturday the weather had moderated. Jason joined me as we walked back to the far depths of Flat Rock to wrap up the first round of data collection. As part of his work with the Forest Service he had conducted similar botanical monitoring. As we worked through the first few plots I was happy to find that his estimates of coverage for each species matched mine. It was an affirmation that my work was accurate. Also, no matter how much of a loner someone imagines themselves to be, the company of a friend is always welcome when engaged in tedious activities.

About halfway through the day's work we sat down on the limestone slab where Meadow and I had built our fire and ate a picnic lunch of Subway sandwiches. I noticed that a sign now hung by a nail on a small winged elm behind the slab. It reminded visitors that many things are forbidden in state natural areas. Among them was alcohol, pets off leash, camping, and fires. I looked down at the spent embers from the fire I had built a couple of weeks before.

"Well, that sign was obviously hung in response to my little fire," I said. "It took some doing to find this spot."

"Yeah, it's definitely off the beaten path," replied Jason. "Have you gotten any emails from the state about it? Those nails with that pink ribbon are hard to miss. It's obvious somebody is doing research here."

"No. But they wouldn't necessarily know it was me."

"Well, I'd be careful in the future. I think they're pretty strict."

At this moment I noticed a commotion in the woods as a camouflaged figure broke through the tangled briars and cedars and walked into the open glade. He approached us, taking note of the moisture probe and PVC plot on the ground.

"Looks like you are doing some botanical monitoring," said the mysterious figure.

"Yeah, trying to figure what all these undeveloped sprouts are," I said.

"Carl Himebaugh," said the man in camouflage, extending his hand. "I'm the state botanist, just coming out to see what's in bloom."

"Billy Plant," I replied, with a handshake. "Maybe you can help me figure out what some of this stuff is. It's hard to tell sometimes without the flower. For instance, are all those onions?" I asked pointing out to the little streamside meadow that lay before us.

"Some are, but most of those are sunnybells," said Carl.

Sunnybells (*Schoenolirion croceum*) are an endangered species found in wet zones around cedar glades. In the 1980's a section of the 840 bypass through Rutherford County had to be rerouted to preserve a population of them. About this time my dog came running out of the woods, crossing the stream where the sunnybells grew, kicking up mud as she tore across the glade. I wanted to crawl under a rock. But Carl never mentioned it.

Sometimes in our quest to understand something we play a small role in its temporary destruction. The footprints of a fast dog were far less detrimental to the sunnybells and the soil that supports these fragile plants than the grinding feet of two researchers bent over a plot attempting to understand why they grow where they do. To explore a place inevitably diminishes it ever so slightly despite our best intentions. The cedar glades formed over millennia without having to develop a response to the heavy footsteps of man. But only by entering special places can we tell the world about what is there and hopefully make a case for protecting the landscape against the deeper, more permanent scars of development and agriculture.

*　　　　*　　　　*

In late April the periodic cicadas emerged from the ground. They are white after they first molt but quickly become black with red eyes. The cicadas were all the buzz when they first arrived. Local news shows and magazines ran features describing what would happen if your dog or cat ate too many cicadas. Then, on the next page or television segment, they would interview guest chefs offering cicada recipes for human consumption. In their second week the cicadas took a hit from Mother Nature as the weather once again turned off unseasonably cold and drizzly, with highs only in the lower fifties for over a week by the beginning of May. Then as so often happens summer abruptly arrived to stay.

By the second full week of May highs were hitting ninety degrees. Once again I returned to the glades with my moisture probe and plots for the second round of data collection. By this time I had taken on a new project, the result of which you are holding in your hands. Beginning in March I had started going into the field collecting specimens to identify and mount for a flowering plant class I was taking which was taught by the aforementioned elder botanist. As I discovered the flowers in bloom I would photograph each one with a little point-and-shoot Canon camera. This pastime became addictive. Every four or five days I felt obligated to get out in the field and see what was ready to be photographed. Through the photographs I was able to change my understanding of the flora of the hills, valleys, and cedar glades of Tennessee from a mass of colorful wildflowers to a knowledge of distinct individuals, each with their own story to tell of life cycles and medicinal or economic properties.

When I uploaded the photos onto Flickr, the online photo sharing site, I would type in a few comments about the species and always included the date and location where it was photographed, sort of like a digital herbarium. Perhaps in time this will help future botanists better understand how distribution and habitats have changed as climate change slowly shifts our seasons and precipitation patterns.

The May data collection went smoothly but tediously. Having gone out sampling now on various projects with professionals from several government agencies and the private sector I have never met anyone who did not want to get the sampling done as quickly as possible. On one occasion a National Park Service employee told me that he had rather be out fourteen hours a day fighting forest fires than conducting botanical monitoring. Botanical sampling is work that demands constant attention, often kneeling down in cramped positions in wet, hot or otherwise uncomfortable conditions.

Many of the plants that had been unidentifiable a few weeks before had blossomed into their prime. Nashville breadroot (*Pediomelum subacaule*), Missouri evening primrose (*Oenothera macrocarpa*) and the delicate glade sandwort (*Arenaria patula*) were among the showy flowers adding an understated beauty to the glades. Still there were sprouts of plants that would not be in their full glory until late summer that made identifying them difficult.

*   *   *

With the close of the Spring semester and the end of my job teaching biology labs at the university I anxiously awaited the call from Stones River National Battlefield to find out the start date for my seasonal position as a biological technician. But the call was slow in coming. Having some free time I threw my Navy seabag in the back of my truck and headed off to the mountains with my dog. Murfreesboro, the center of the universe when it comes to cedar glades, is as flat as the Plains (incidentally a few species that are rare in Tennessee, found only in the glades, occur in abundance in the prairie states...these are called disjuncts). Occasionally I feel the need to walk up mountains and stand before roaring sheets of falling water.

I had also self published a book called *Mother Earth & Other Pretty Girls* during the winter. Being a book that integrates stories about post traumatic stress disorder, Mount St. Helens, and alcoholism with overly emotional or esoteric poems and photographs, I imagined it would have already been a best seller. Far from it. The only book review it had received thus far was in Murfreesboro's little entertainment monthly and began with the lines, "This book is so bad it killed my grandfather. Not really, but sort of." My plan was to go to Ashville, North Carolina, one of my favorite mountain towns, and try to sell copies to independent book stores and possibly get the book reviewed in a local paper.

After a pleasant drive up the Blue Ridge Parkway and a night spent camping by Lake Powahaton just outside Ashville I awoke the next day and drove into town to peddle my book. I sold a copy to a book store owner who told me that if he thought it was right for his store he would order more copies. Years later I've never heard back from him. I stopped by the office of an alternative weekly newspaper offering a copy for a review. They kept the copy but I never got the review. Disheartened I had lunch at the Ashville Brewing Company and sat on the patio with my dog. I asked my waitress where her favorite place was to go hiking in the area but she didn't have one. Being in a band that toured the festival circuit she said she never had time. I gave her a copy of my book also.

After leaving Ashville I drove into the Nantahala and Pisgah National Forests. I took a short hike to a pretty waterfall under the cool canopy of yellow poplars, oaks and a few surviving hemlocks. Leaving there I wound my way around several miles of forest service roads, jarring across the coarse brown gravel and mud. Near Tellico I hiked a small section of the Appalachian Trail until dark, photographing false Solomon's seal, speckled wood lily, wood betony, and a handful of other upland wildflowers along the way.

Once again on the back roads, I listened to a preacher on the radio talking about the creation story from Genesis. He advised against environmental stewardship under the pretense that God had given Man dominion over the earth and that to extol the virtues of our planet too emphatically would take away from the glory of God. I have always imagined God would appreciate His guests cleaning up the mess we have made while imploring future generations to take better care of our earthly home.

Late that evening I pitched camp beside a rather large, tumbling creek that was strewn with boulders. I awoke to the invigorating sound of running water. I ate a breakfast of peanut butter and Triscuits washed down by instant coffee.

A few miles up the road I visited Joyce Kilmer Memorial Grove. You've likely heard Kilmer's most famous poem even if you don't recognize his name: *I think that I shall never see/ A poem lovely as a tree...*Kilmer was killed on a reconnaissance mission in World War I. Afterwards his American Legion post petitioned to have a forest named in his honor. The veterans in conjunction with the government found a stand of large trees that had only been spared because the lumber company that owned it had recently gone bankrupt. Walking through Joyce Kilmer Memorial Grove one encounters two hundred foot tall yellow poplars up to twenty feet in circumference. That is a big tree by deciduous standards.

Nowadays, perhaps more striking to the eye than even the large poplar trees, is the utter annihilation of the hemlocks in the area. They are all dead. When I had visited a year before the hemlocks had been standing as massive dead snags. In the intervening year the forest service had decided the standing dead trees posed a hazard to visitors so they had gone in with explosives and blown the dead trees down. The rational was that by blowing up the trees instead of cutting them with chainsaws the forest would look like it was destroyed by windshear. But it was a much smaller force than wind or chainsaws that actually killed all the hemlocks in western North Carolina. The culprit is a little insect the size of a pin head called the wooly adelgid, accidentally introduced from east Asia. It is an exotic species that has already destroye most of the hemlocks in Shenandoah National Park and the mountains of North Carolina and is now infesting forests as far west as the Big South Fork area in Tennessee.

My stay in the mountains was a pleasant one but I was running out of money so I had to get back to Murfreesboro. Following a brief stint helping out a landscaper I was finally able to start at the battlefield by mid-June. The rest of the summer found me cutting, pulling, and spraying herbicide on invasive species such as Chinese privet, bush honeysuckle, Johnson grass, and Japanese stilt grass in the employment of the Unites States government.

This being a book about cedar glades you may wonder why I have included this long detour about camping in the mountains and working at the battlefield. The reasons are two-fold. One, every place I have talked about in the last few paragraphs is a special place that people have seen fit to protect. That protection begins at the personal level with the vision of one or a few individuals who formulate a plan then seek the help of government agencies to implement it. Some in the recent political climate choose to give any form of government action a negative connotation. That is simply wrong minded. Some of the greatest visionary successes we have achieved as a nation, good for all people across all economic levels, have been preserving our natural heritage in the form of scenic landscapes and unique habitats and protecting endangered species, and ensuring clean air and clean water. Secondly, poor planning and a lack of adequate regulation is responsible for the invasive species now taking over our landscape and killing off many of our most beautiful forest trees such as the hemlock, the ash (emerald ash borer from Asia), chestnut (a fungus from China or Japan), and soon to be walnut (likely an exotic fungus).

*   *   *

Summer grew oppressively hot with the coming of July. Working at the battlefield I was able to settle into a comfortable routine: up every morning by 5:30 a.m. to make a pot of coffee, at work at 7:00, fight the battle against invasive plants until 3:30, take my dog for a walk and spend time with friends in the evening, then early to bed. Some nights we would build an unnecessary but psychologically pleasing fire in the back yard and sit around it passing a guitar while our dogs lazed around in the shadows lulled by the sounds of insects and the smells wafting on the night air. On other nights I had band practice. Our little four piece ensemble played an eclectic set of songs by the Rolling Stones, Billy Shaver, John Prine and others with a few of my originals sprinkled in. We played several shows that summer. Although I have played guitar since I was thirteen years old this was the first time I had ever been onstage. I immensely enjoyed the interaction with an engaged audience. Maybe it was an unconventional way to live for a man pushing forty. But it was a lifestyle I had often dreamed about in my former life as an officer in the Navy, deployed at sea for months on end. The vision of such a free-wheeling life carried me through some scary times in Baghdad as our compound was being assaulted by exploding (but poorly aimed) rockets and mortars.

Throughout the summer I made at least weekly trips to Flat Rock, usually with no other companion than my dog. As summer progressed the glades grew hotter and dryer. The wildflowers of May shriveled, following the lead of the plants of early spring which were already gone. By the time I collected data for August the glades were scantily clad in a sparse combination of herbaceous plants and grasses: Heliotrope (*Heliotrop tinellum*), Gattinger's prairie clover (*Dalea gattingeri*), puny looking glade petunias (*Ruellia carolinensis*), little blue stem grass (*Schyzachrium scoparum*), and the aptly named poverty dropseed (*Sporobolus vaginiflorus*). The most striking cedar glade plant of late summer, usually found at the glade edge growing under the native glade privet is the blackberry lily (*Belamcanda chinensis*). As its specific epithet implies this plant is an exotic, though not an invasive species. It has escaped from cultivation where it has been planted for many years for its red and yellow daylily-like blooms.

As I've said by late summer cedar glades are hot, dry places. During the August period of data collecting there was no water. But late summer still offered many opportunities for photographing plants in the glades. On my frequent trips I photographed the thoroughworts (*Eupatorium altissima*), goldenrods (*Solidago* spp.) and the abundant, lavender flowered false pennyroyal (*Isanthus brachiatus*). One Saturday morning I awoke achy headed after a night of celebrating the joys of barley and hops. But outside my window the sunshine was warm and beckoning. I picked up my guitar and wrote a little song about my efforts to chronicle life in the glades. I sing it to the tune of "Wildwood Flower".

I woke up this morning with an aching in my head
I guess I'd been out drinking when I should have been in bed
Coffee gives me jitters all through morning's dreary hour
But the sunshine at my window makes me want to find wildflowers.

Eight hundred lonely acres and my dog at my side
A little Canon camera and Hemmerly's field guide
We walk across the glades with the sun beating down
There's not a drop of water and there's no one else around.

Heliotrope and gumweed black berry lilies in bloom.
We walk past a Baptisia that was blooming in June.

Nostoc crunches loudly on the limestone at my feet.
Gattinger's prairie clover with its perfume so sweet.

Ordivician fossils lay embedded in the rocks.
Red cedars in the woods are twined with Smilax bonanox.
My dog runs through the trees where coyotes roam free.
They nip her in the haunches
And chase her back to me.

My look at the sky the August sun is getting low.
I've enjoyed the wildflowers here but now its time to go.
I'll be back again to see what can be found
But I'm ready for a cold one so I'm headed back to town.

*　　　　　*　　　　　*

Autumn is generally a warm, dry season in Tennessee. Such was the case during the year of my study. In mid-October I went out collecting data and found that once again people had returned to the cedar glades. Like me they had come out to enjoy the warm, dry air and cobalt blue skies. The golden rods and grasses had gone to seed, nearly completing the growing season. But there were still some plants to photograph, mostly asters. I am haunted by asters. Many of them being very similar, they are difficult to identify unless taken into the lab and examined under a dissecting scope. And frankly I was tired of identifying plants. I had identified many dozens over the past few months and the asters presented a special challenge due to their seeming lack of morphological diversity. Still, using Ronald Jones' *Plant Life of Kentucky* I was able to key out a couple of the most prevalent cedar glade species for this field guide. I also finally photographically referenced all of my transects. I pulled out the tape between the nails and took a picture from eye level. For me that is roughly 1.75 meters high. It should have been done earlier in the process but doing something a little late is better than not doing it at all.

In late October I once again felt the need to go to higher ground. No longer working at the battlefield and my teaching schedule being undemanding, I had been drinking too much throughout the fall. This happens late in the year as the days grow shorter. I know the dark days of November and December are approaching soon and with them my mood grows darker as well. I suppose this is true for everyone but sometimes November and December find me feeling hopeless. It is called seasonal affective disorder (SAD). It affects different people in different ways, from a loss of appetite, to substance abuse, to withdrawal from society. I am fortunate that my case is not severe. But often at these times I want to be alone.

With such feelings looming on the horizon my dog and I took off for a camping trip to South Cumberland State Park. I pitched my tent in a little campground near the Alum Gap Overlook. On a chilly day in late autumn I walked down to Greeter Falls. The thirty-five foot waterfall gushed more heavily than usual in the season due to recent rains. I walked down the creek to lower Broadtree Falls. Some hydrangea hung on stubbornly to the growing season, tucked into a large crevice behind the falling water. I took a picture of them and ate lunch. Then I hiked up out of the gorge and sat at Alum Gap until sunset. This is one of the prettiest spots in Tennessee. The few blazing stars that lingered there danced on the stiff breeze that blew across the sandstone bluffs. From this perch I could look down on the backs of turkey vultures as they rode air currents high above the gorge but still below my feet. The dissected fingers of the plateau stretched off into the distance under the bluest of skies, which gently faded as the sun began to set behind the ridge.

That night I was alone in the campground. After a dinner of ranch beans I sat by the small fire until its load of fuel flickered one last time and died down. I crawled into my sleeping bag and my dog curled up in her familiar corner of the tent. I read a little of Waade Davis' *The Serpent and the Rainbow* before drifting off to dreamless sleep, huddled in my sleeping bag against the sharply falling temperatures. The next morning I awoke to a heavy sheet of frost across my tent. With this frost came the end of the growing season I had followed so closely.

*   *   *

I returned to the glades in late November on a melancholy day under a blanket of gray winter clouds. Everything looked so different than it had during the spring and summer. Water flowed freely from the little spring, flooding the meadow in front of me as I sat on my familiar limestone slab. The frost-burnt tops of onions and grasses mingled with the brown, dead stems of bluets and St. John's wort. But even so the little basal rosettes of Nashville mustard, which wouldl give the glades a pleasing and aromatic blush of white and yellow in late February, had already sprouted. Sitting there I wrote a very bad poem:

Rocky open place, barren and dry,
Reflecting the seasons of my heart.
My eyes look out on the wasted vegetation
That follows the forgiving frost.
Flattened towers of rocks I've stacked
While drinking here with friends
Lay fallen and scattered on the big rock where I sit.
And this rock where I sit
Has been my home for a year.
But it's fissured and flaking face shows
That even rocks don't last forever.
But as the glades brown over in death and dormancy
The waters return.
The little spring runs again between the fissures
And the mechanism for rebirth
Bubbles up from below.
In the midst of the end
The new beginning has already been foretold.
Can you see it?
Across the expanse of rock and earth, can you see it?

I see this poem as being in the tradition of the Romantic poets. I am a Transcendentalist at heart. Still, I think both schools of thought benefit from a healthy dose of reality. Do not speak to me of the benevolence of Nature until you have heard the shrieking rabbit in the talons of a hawk. I was finally jolted out of my funk when a little cattle dog that had began following me on my trips through the glades ran in out of nowhere and jumped up hitting me square in the face, knocking me over. I pushed her off but she jumped on me again. I sat content, rubbing her behind the ears, which is what she wanted. My dog sniffed the ground at the edge of the woods. I think she was glad the little cattle dog had finally decided to leave her alone.

Financially times were very tight that winter. Throughout the fall people had taken to setting up camps in public parks and plazas across the United States in what was known as the Occupy movement. I believe that initially these "occupiers" did represent a broad cross section of Americans fed up with corporations controlling the vast majority of wealth, and by default, many politicians. They had reasonable spokespeople who made valid points. But by December the character of this movement had devolved into one of protests by contrarians more idealistic than motivated by constructive dialogue. My own finances had been precarious for some time but another round of car trouble finally caused me to have to radically change my lifestyle. I sat at home more often.

During this time I had a recurring dream where I was walking along the edge of a beautifully verdant waterfall. Suddenly I'd find myself walking along a precarious rocky ledge. I'd have to climb across obstacles with shaking knees and hug the rock for the narrowest of hand and footholds.

*   *   *

The morning was very cold. A heavy frost lay white on the grass and thick on windshields. But the lack of cloud cover that had allowed all the heat to dissipate overnight meant the air would warm quickly as the sun climbed in the sky. Meadow and I walked down the long trail knowing we had work to do. I would pull the transects, measure the soil moisture and depth, and record whatever plants were still green in each of the plots. By now there were few plants to record. Just a few basal rosettes of Nashville mustard, fleabane, Nostoc, some Spirogyra, and some iridescent green glade moss.

The work went well. We were done by half past one. To celebrate we sat on the big slab of limestone and shared a bottle of Belgian-style beer. By afternoon the temperature had climbed to forty-five degrees. With no wind it felt very comfortable sitting in the sun. As a last task we pulled up the nails that marked the end points on the completed transects. I made sure to gather up all the loose pieces of florescent pink flagging that had by now grown faded and brittle from a year under the piercing ultraviolet rays of the sun in the cedar glades.

*   *   *

Months have faded into years since Jason and I drove those first nails into the ground on that cold January day. In May 2012 I defended my thesis and earned my Master's degree. That summer Tennessee experienced a heat wave that broke all records. One Friday after cutting privet and honeysuckle at the battlefield I prepared the treatment record at the end of the day and recorded a temperature of 110° F. I can only imagine what the mercury had risen to in the glades. But extreme drought and record temperatures had not destroyed the tough little glade plants. On an afternoon trip to Flat Rock I found fleeting green remains of Gattinger's prairie clover, limestone fame flower, and St. John's wort interspersed among the rocky substrate while little bluestem and poverty dropseed continued to thrive.

After earning my Master's degree I returned to surveying which is the work I did before going to graduate school. I eventually took a job as a vegetation management contractor for a regional utility company. Both of these jobs inevitably have negative impacts on the landscapes I would love to preserve. But I do not lament development or progress so long as these two forces are approached with sustainability and the overall health of the landscape in mind. Both are necessary for job creation and to improve the overall quality of life for the majority of the people. This field guide is in some ways my atonement for the habitat destruction I have been part of. It is also my way of remembering a specific time in my life when I felt free, when I had the luxury to focus on gaining knowledge more than money. I took pictures along the way. We have to take pictures because we forget what we have seen or turn the past into what it was not...Memory is a clever creator. For me this book is a document of that time.

Just as I have tried to tell the story of a year spent in the Tennessee cedar glades in these pages the glades themselves will bring it to life much more eloquently in the coming seasons. Walk through a cedar glade in late April and you will find delicate wildflowers that dance on a breeze, ready to reveal to the patient observer how they have come to thrive in a harsh landscape found nowhere else on Earth.

Botanical monitoring equipment.

A Brief History of Plants and Botanical Research

"The weight of a petal has changed the face of the world and made it ours."
-Loren Eiseley

Respect plants. Without them humans would not exist. Everything that changed a chunk of rock surrounded by vaporous gases into a hospitable planetary home can be traced back to the rise of photosynthetic organisms some 3.5 billion years ago. The first proto-plants to produce energy by harnessing the power of the sun were the cyanobacteria, formerly called blue-green algae.

Cyanobacteria get their name and photosynthetic ability from the pigment protein complex phycocyanin. This pigment absorbs light energy from the sun which reacts with atmospheric carbon dioxide (CO_2). Carbons from pycocyanin use free electrons from water molecules to drive the photosynthetic pathway which ultimately yields carbohydrates and molecular oxygen. This release of oxygen is the key to life on Earth as we know it today. Prior to the evolution of cyanobacteria Earth had a reducing atmosphere, the critical component of which was CO_2. This is due to the fact that oxygen is unstable, quickly reacting with other elements to take on a new molecular identity. Such an environment proved beneficial for an organism which could create food from sunlight and an abundance of CO_2 in the atmosphere. As the cyanobacteria flourished more and more oxygen was pumped into Earth's atmosphere until ultimately molecular oxygen (O_2) replaced CO_2 as the second most abundant gas behind nitrogen (NO_2). The nature of life took a sudden and dramatic change from anaerobic to aerobic. Today most anaerobic organisms are bacteria and are tucked away into the darkest corners of life on the planet.

Though very old, cyanobacteria are still thriving. They account for 20% to 30% of the photosynthetic productivity on Earth, which means they are still contributing greatly to the oxygen you and I so inherently depend on. So how does this excursion through the rise of the first proto-plants tie in with the modest scope of this book? A cyanobacteria is present, even abundant in cedar glades. When you hear that crunch beneath your feet as you walk across the sun-scorched limestone and look down to see what looks like a charred sheet of crumpled paper you have found it. It is called *Nostoc*. When touched by the least bit of precipitation it reconstitutes itself into a dark yellow-green gelatinous mass no more attractive than its common name, witches' butter.

Next we move on to the rise of green algae which was the immediate forbear to the evolution of true plants. The term endosymbiosis literally means "living together within" though not in that order. The term can be applied to the bacteria that line the human gut and facilitate our digestion. At its most basic level endosymbiosis began when an organism ingested an aerobic bacteria through the process of endophagocytosis. Imagine an amoeba wrapping its pseudopodia around a smaller organism and you get the idea of how it might have happened. According to endosymbiont theory, chloroplasts, which we associate with photosynthesis and the green color in plants, are the evolutionary product of cyanobacteria becoming incorporated in the cells of more complex organisms. This notion was arrived at by observing the morphological similarities between chloroplasts and cyanobacteria. The fact chloroplasts, like mitochondria which drive energy production in animal cells, have their own DNA separate from that of the host organism, and the fact that they have membranes unlike any other organelles within plant cells, is a good clue that the endosymbiotic theory of chloroplasts is probably correct.

Now fast forward through time....a long time, to the rise of more complex plants. Past the lichens and mosses. Fast forward to the Paleozoic Era, during the Devonian and Carboniferous periods 360 million years ago when amphibians were beginning to diversify and the first reptiles begin to appear in the fossil record. The vegetation that supported these creatures consisted of giant ferns and *Equisetum* (scouring rush), collectively known as pteridophytes. These were the first vascular plants which means they had a mechanism for internally transporting nutrient bearing fluids and water. Plant reproduction occurred through the production of spores which would drop to Earth and vegetate into a gametophyte which bore sperm and eggs for sexual reproduction. The plants were confined to moist environments, barely emerging past the shores of a body of water. Why? The sperm needed a medium in which to reach the egg. Once fertilized, the egg, now called a zygote, would undergo rapid cell division, obliterating the ephemeral gametophyte, replacing it with a new sporophyte generation which in turn produced new spores to continue the process. When you look at a fern or a pine tree you are looking at the sporophyte generation. One more interesting note, the term Carboniferous alludes to the fact that the carbon based oil and coal we use today was formed by the geologic transformation of the ferns and other early vascular plants which once dominated the world.

Now fast forward again, stopping about 200 million years ago during the

Mesozoic Era, at the tail end of the Triassic and on the cusp of the Jurassic periods. The terrestrial world was dominated by dinosaurs, some of them very large though the biggest were still yet to come. The earth's flora was dominated by the gymnosperms ("naked seed"), many of which, like the dinosaurs were of magnificent size. By now plants had evolved a new mechanism for their deployment and eventual domination of the land: it is called the seed. The evolution of seeds allowed a tiny encapsulated plant to ride the wind farther and farther away from the parent plant, reaching new frontiers. The seed was packed with just enough energy for the tiny plant to complete the journey and germinate if it was blown to a location with favorable light, soil, and moisture. Today the coast redwoods and giant Sequoias, ginkgo trees of city parks and a few remaining cycad species bear witness to these ancient forests of so long ago when we must stretch our imagination to its limits to truly envision what life would have looked like in such a landscape. In the glades the Eastern redcedar (*Juniperus virginiana*) is the only representative of the conifers which once dominated the Earth.

Our travel through time has taken us a long way since the rise of the cyanobacteria in the toxic, yellow and brown, carbon dioxide and hydrogen sulfide rich unpleasantness of Earth's earliest days. Even after the rise of the land plants the Earth wore a vegetative cloak of a rather dull green. Then, about 130 to 100 million years ago, give or take a few million years (standards of exactness are pretty liberal in geologic time) a new kind of plant germinated for the first time, poking its apex shoot through the soil to change the Earth once again.

The angiosperms ("vessel seed" due to the fact that the seed is inside a nutritive or protective coating) came upon the scene dramatically, abruptly enough for even Charles Darwin to call their explosive appearance in the fossil record a "perplexing phenomenon" and an "abominal mystery." The key to the success of flowering plants was the amount of nutrition that was packed into each seed. These new nutritive powerhouses, which allowed the seed to travel and remain dormant for extended periods until proper conditions for germination occurred are the result of a process called double fertilization. In gymnosperms two sperms are released when a pollen grain makes contact with an egg. One sperm fertilizes the egg but the other disintegrates. Pollen from flowering plants also releases two sperms, one of which fertilizes the egg. The other sperm combines with other nuclei within the ovary creating a nutrient rich endosperm. It is this endosperm that we are eating when we crack open a walnut or grind wheat into flour.

The first flowering plants relied on wind for seed dispersal as most still do today. Tufts of hairs called a pappus catch the wind, lifting the seed from the parent plant to be deposited farther afield. Just think of blowing that fluffy dandelion head and you get the idea. Competition prompted other angiosperms to devise more intricate means of pollination and seed dispersal. Many plants evolved fragrant flowers in brilliant colors to attract the insects present at the time. These insects, mostly beetles, would visit the flowers to feed on sweet nectars while becoming coated in pollen during the feeding process. As this pollen was spread between more and more members of the species genetic diversity increased. In order to compete with other flowers for these conjugal visits some flowers became more and more specialized over time. In order to receive these nectar rewards insects too became more specialized in a process called co-evolution. Perhaps the process has gotten a little out of hand because today there are orchids with flower morphologies that mimic the female abdomen of their insect suitors while others have achieved such a degree of specialization that they are totally dependent on one species of insect for pollination. If you visit enough cedar glades you will eventually encounter yuccas near the margins of some. These plants with their flamboyant spike of white flowers are totally dependent on the yucca moth for pollination and the moth will not pollinate or associate with any other plant.

As the earth's climate changed so too did its landscape. While the dinosaurs disappeared and the first rat sized mammals and feathered birds stepped into the bright sunlight of existence the giant gymnosperm forests receded into the higher altitudes as grasses swept across the plains and savannahs of a new era. Over time the mammals grew bigger, faster, ultimately becoming the dominant land animals. The first human ancestors climbed down from the forest canopy and wandered into the expansive vistas of the grassy savannahs. As she is wont to do Nature experimented with several forms, arising from dust in response to unfilled niches, then quietly sinking back into Earth after untold millennia, existing today only as clues in the fossil record, an evolutionary dead end.

Modern humans are pretty lucky to have escaped a similar fate. Perhaps it was the large nimble brain that saved us, overcoming the limitations of our relatively weak bodies. With no claws and no ominous incisors, the human form seems to be made for little more than running away. But that brain gave us leverage. What we lacked in large teeth we made up for by making tools of stone to do our piercing and shredding for us. With these tools early humans could hunt the plentiful beasts of the field. And eventually, in a breakthrough of esoteric thought, humans realized they too could feed off the abundant grasses which nourished their animal prey. Man began to grow and refine the energy rich ancestor of wheat. He no longer had to move with herds of animals. With the abundance of food created by early agriculture humans had enough food to feed animals as well as themselves. With the domestication of these animals man's food resources were somewhat secure. Humans now had time to pursue broader realms of thought. Man created governments to administer the production and distribution of food. He created art and myths to explain the things in his ever expanding world that seemed unexplainable.

I don't think I can overstate the case for plants being the basis of civilization. Without discovering how to grow and control a dependable food source, namely wheat and rice, there would be no mathematics, no philosophy, no great buildings, trade routes, or empires.

After the rise of civilization and the early intellectual triumphs of the Egyptians, Greeks, Romans, and Chinese much of the western world sank into a deep darkness, based on a power struggle between the Church and anyone whom it felt challenged its total dominance by daring to investigate observable phenomena rather than blindly following doctrine handed down from the seat of power. Many of the great advances of early civilizations were lost while others went unimproved on for 1,000 years. It wasn't until the late sixteenth century that humans once again began to observe the world around them and publish their findings, many still under the fear of persecution from Church or state powers (which were often one and the same). But reason flourished and eventually found its place in the new middle class. New fields of study were born as chemistry overcame alchemy, botany overcame witchcraft, and philosophical reasoning once again overcame doctrine and dogma.

By the mid-1500's Europeans began exploring a new world in the western hemisphere where they encountered Native Americans whose ancestors had crossed a land bridge from Asia 10,000 years before. These aboriginal cultures had developed many types of food that were unknown to the outside world. Spanish and Portuguese explorers found the Incas growing hundreds of varieties of potatoes and tomatoes in all colors and sizes and eating curiously hot little fruits the explorers called "peppers" after the pungent spice they had been seeking in the first place though the two plants are not related at all. Most importantly they found an energy packed grain that the Indians had cultivated from a much speculated ancestor plant which has possibly been extinct for thousands of years. How the Aztecs manipulated the tiny seed head of a local grass into the most important food crop of the modern world is still a bit murky. But there is no doubt that maize/corn is likely the most highly engineered food plant to ever arise from human endeavor.

By the eighteenth century the observed knowledge of the world was expanding at an unprecedented rate. Botanists, geologists, and scientific artists joined exploring parties at sea and over land. New disciplines arose within disciplines. Far from the frontiers of exploration a new kind of scientist arose. He or she didn't hold a government commission to explore lands on behalf of the Crown. Instead these amateur naturalists lived a rather mundane life as rector of a parish church, doctor, or some position within the merchant class. What these early scientists did have was an ever increasing base of published information to draw upon and time to walk around the countryside, observing birds, plants, fossils, weather and the like. Which brings us back around to the study of cedar glades.

A History of Cedar Glade Research

I'll be frank: cedar glades aren't much too look at if your primary interest is growing crops, grazing animals, or exploiting forests. They are rocky open expanses that support rather delicate herbaceous plants and a few small, dispersed stands of grasses. They are surrounded by woods made up of redcedar trees and stunted oaks, elm, and hickory trees. About the only industry they have ever fostered was the rise of pencil making in middle Tennessee in the late 1800's. But cedar glades are interesting and beautiful in their own quiet way. Still, it took years for people to realize this. Not until 1851 were cedar glades mentioned in the scientific literature when J.M. Safford, writing in the *American Journal of Science and Arts*, published a geologic map of Tennessee in which cedar glades were depicted by small clusters of cedar trees. In 1874, writing with J.B. Killebrew, Safford describes the glades as redcedar forests interspersed with rocky openings.

The first person to study the botany of cedar glades was Augustin Gattinger, a German immigrant working as a physician in Nashville after the Civil War. His 1887 self-published work, *The Tennessee Flora With Special Reference to the Flora of Nashville* was a milestone in understanding the floristics of cedar glades in that it was the first cataloging of of the cedar glade flora, which of course is what the glades are known for today. In *Tennessee Flora* and his 1901 work *The Flora of Tennessee and a Philosophy of Botany*, as well as in a series of letters Gattinger describes scenes that a visitor to the cedar glades will still find today. He writes of making his way through cedar woods where, "The cedars are always closely set, it is a vexatious and ungrateful task to penetrate such thickets." But the reward is to emerge from these dark, thorny tangles to stand within the open glade and observe "with dazzling vividness the crumbling limestone flats, overspread with the rosy *Sedum pulchellum* and carmine-flowered *Talinum*, or the golden stars of the *Opuntia Raffinequii*."*

After Gattinger more botanists began to study cedar glades. Harper, Picklesimer, and Freeman all made valuable contributions to the study of cedar glade ecosystems during the first forty years of the twentieth century but the queen of the cedar glades is Dr. Elsie Quarterman of Vanderbilt University. Between 1947 and 1993 she published numerous papers which greatly expanded the scientific knowledge of glade development and flora. She was the first to describe differing habitats within glades based on soil depth, plant communities present, and light intensity. Furthermore, working with Dr. H.R.

DeSelm, she clarified the distinction between cedar glades and barrens, naturally occurring grassy areas on Tennessee's Highland Rim. No history of cedar glade research would be complete without mentioning two of Dr. Quarterman's acolytes. Drs. Carol and Jerry Baskin of the University of Kentucky have researched and written dozens of articles on all aspects of cedar glade ecosystems. Their work has added extensively to our knowledge of everything from seed germination to different ways various cedar glade endemics have adapted to the harsh glade environment. In a piece I have drawn on heavily for this essay they review the scientific literature to give a complete history of the use of the term "cedar glades" and what that has meant over the years.

Part of the reward of studying cedar glades is that it is a small community of scientists with close roots to the past. I conducted my Master's research under the mentorship of Dr. Jeffrey Walck who in turn earned his Ph.D. while working with the Baskins who, as mentioned above studied under Dr. Quarterman. Dr. Thomas Hemmerly, another of Dr. Quarterman's students, wrote *Wildflowers of the Central South* (1990) which first introduced me to the geology and flora of cedar glades and awakened my fascination with this aspect of the natural world. Twenty years later I had the opportunity to take his medicinal plants class at Middle Tennessee State University. He, along with Dr. Walck and Dr. Kim Sadler, was instrumental in the establishment of the Center for Cedar Glades Studies at MTSU. The CCGS web site is a cornucopia of information about cedar glades, offering everything from photographs of plants, to lesson materials for teachers, to an updated bibliography of cedar glade research in the published literature.

Now it is your turn. Head out into the field with a camera and a friend. Enjoy the quiet beauty of the glades. Contemplate the environmental pressures under which these little plants strive to germinate, grow, and produce the seeds of the next generation. Perhaps William Blake said it best: *To see a World in a Grain of Sand/And a Heaven in a Wild Flower/Hold Infinity in the palm of your hand/And Eternity in an Hour.*

*In modern nomenclature the first letter of the genus is always capitalized and that of the species is lowercase. Up until the mid-1900's specific epithets derived from the name of a person or a geographic location were often capitalized. Botanists were the last to let go of this spelling convention.

Cedar Glade Ecology

Cedar glades (or limestone glades) are seemingly incongruous open areas surrounded by a forest of shrubs, redcedar, elms, oaks and hickories. Glades have been found in various locations throughout the south central United States, namely northern Georgia and Alabama, southern Kentucky, and the Ozarks of Missouri and Arkansas. However they reach their peak development and boast the highest number of glade endemic species in Rutherford, Wilson, and Davidson Counties in Tennessee. Glades are also present in Bedford, Cannon, Giles, Marshall, Maury, and Williamson Counties in middle Tennessee.

Cedar glades have been described as having six distinct zonation patterns ranging from exposed sheets of Lebanon or Ridley limestone at the center of the glade to gravelly areas with thin soils, to a ring of shrubs and redcedars at the glade edge, which gives way to oak-hickory forests. The gravelly areas of zone two are populated by small herbaceous plants which is where the majority of glade endemics occur. Radiating outward toward the glade edge soils become deeper with grasses as the dominant flora.

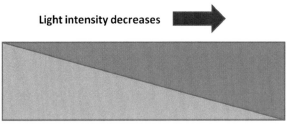

Light intensity decreases

Bare rock to gravel
0 – 5 cm soil depth

Grasses to cedar forest
5 – 20 cm soil depth

Cedar glades are very hot and dry, up to twenty degrees warmer than the surrounding woods in summer, then becoming saturated in the winter and early spring. In these harsh conditions a unique array of endemic plants have adapted to fill niches found nowhere else on Earth. Hence many of these plants are so rare as to be imperiled by the rapid development that has occurred in middle Tennessee during the last thirty years.

Flower and Leaf Morphology

To better identify the wildflowers presented in this guide it is helpful to be able to assess specimens based on the number and structure of individual flower parts as well as the shape and arrangement of the leaves. Though verbal definitions are helpful a picture truly is worth a thousand words. Below are some line drawings that may be helpful to the casual observer.

Parts of a Flower

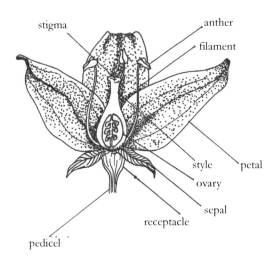

Leaf Arrangement

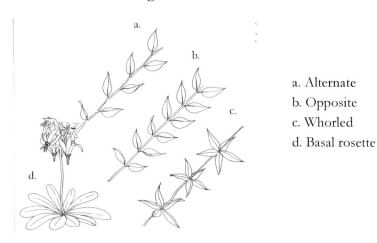

a. Alternate
b. Opposite
c. Whorled
d. Basal rosette

Flower and Leaf Morphology (cont'd)

Leaf Shape

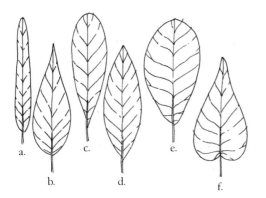

a. Linear
b. Lanceolate
c. Spatulate
d. Elliptic
e. Obovate
f. Cordate

Leaf Margins

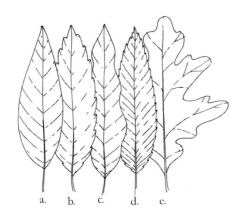

a. Entire
b. Double Serrate
c. Undulate
d. Serrate
e. Lobed

Flower and Leaf Morphology (cont'd)

Leaf Margins

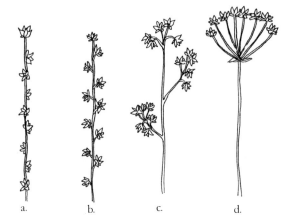

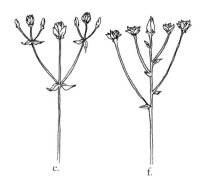

a. Spike
b. Raceme
c. Panicle
d. Umbel
e. Cyme
f. Corymb

Wildflowers Illustrated

Monocots

Alliaceae

Allium cernuum

Nodding Wild Onion - the specific epithet *cernuum* is Latin for "nodding", referring to the way the inflorescence droops over - Ramp (*A. tricoccum*) is a similar species that has flattened leaves; it is considered a culinary delicacy - photographed August 7, 2011 at Flat Rock Cedar Glade, Rutherford County, TN. Leaves basal, linear and flat. Flowers white to pink borne on a bent umbel. July - September. Fruit a triangular capsule.

Amaryllidaceae

Manfreda virginica

False Aloe or Rattlesnake Master - formerly *Agave virginica* - the flowers appear on a shaft rising 3 ft above a basal rosette of succulent leaves - the colorful common name Rattlesnake Master is reference to the sound of the loose seeds in the dried capsules - medicinally False Aloe has been used rubbed on affected area to treat snakebites and the root teas was taken for dropsy, which is an accumulation of fluid in the lungs or other organs and is marked by swelling of the limbs - photographed July 29, 2011 at Flat Rock Cedar Glade, Rutherford County, TN. Leaves are thick and long (to 16") in a basal rosette persisting much of the spring and summer. Flowers are mainly noticeable for their 6 tight tepals and long stamens; female flower parts are marked by 3 small white, translucent petals. June - August.

Araceae

Arisaema dracontium

Green Dragon or Dragonroot - a widespread but somewhat rare plant to stumble across in low, wet woods, it seems as if it should be encountered in the tropics rather than a temperate oak/hickory forest - the root is edible after drying - medicinally used by Native Americans to treat female disorders - photographed July 9, 2011 at Stones River National Battlefield. The single leaf can be divided into a dozen or more leaflets, up to 12" long with acuminate tips. Flowers are tiny and greenish-yellow emerging from a sheath (spathe) at the base of the spadix. May - July. Fruit a dense cluster of red berries.

Nodding Wild Onion

False Aloe

Green Dragon

Araceae

Arisaema triphyllum

Jack-in-the-pulpit - Sometimes called Indian turnip because the peppery tap root formed a part of the Native American diet. Uncooked the calcium oxalate crystals in the will cause a burning sensation on the tongue - photographed on April 21, 2009 at Radnor Lake, Davidson County, TN. Leaves compound and, as the specific epithet indicates, they have 3 ovate leaflets. Flowers inconspicuous borne on a spadix (Jack) surrounded by a ribbed spathe with an overhanging flap on top (the pulpit). April - June. Three subspecies exist with slightly different leaf and flowering structures.

Convallariaceae

Smilacina racemosa

False Solomon's Seal - the species name refers to the structure of the inflorescence - the berries were eaten by Native Americans - photographed May 24, 2011 along Appalachian Trail near Rocky Bald close to Tellico Gap, Nantahala National Forest, NC. Leaves alternate, elliptic, and sessile. Flowers small and white with 3 sepals and 3 nearly identical petals. March - June. Fruit is a red berry.

Hyacinthaceae

Schoenolirion croceum

Sunnybells - a rare wetland indicator species, this plant is found in limestone glade streamside meadows, habitat designated as threatened by NatureServe. In the 1990's this species caused a section of State Route 840 to be rerouted in near Murfreesboro, TN. That population has since been protected and added to the Tennessee State Natural Areas program - photographed on April 21, 2011 at Flat Rock Cedar Glade, Rutherford County, TN. Leaves basal and linear, parallel veined as is typical in most monocots. Flowers with six vibrant yellow tepals. March - April. Fruit is a capsule with shiny black seeds.

Jack-in-the-Pulpit

False Solomon's Seal

Sunnybells

35

Hypoxidaceae

Hypoxis hirsuta

Yellow Stargrass - this perennial is often encountered in the grassy zone around the edge of cedar glades - photographed on April 14, 2011 at Flat Rock Cedar Glade, Rutherford County, TN. Leaves linear and flat, creased along the midrib; as the specific epithet implies the leaves are distinctly hirsute. Flowers yellow, 3 sepals and 3 identical petals. April - May.

Iridaceae

Belamcanda chinensis

Blackberry Lily - though an exotic species, this plant has been in our area long enough that noted 19th century Tennessee botanist Augustin Gattinger believed it to be native - photographed August 7, 2011 at Flat Rock Cedar Glade, Rutherford County, TN. Leaves flat, lanceolate with acuminate tip, up to 18" long. Flowers orange with darker speckles, having 3 sepals and 3 identical petals; ephemeral, lasting for only one day. July - August. Fruit a capsule containing cluster of dark fleshy seeds suggestive of blackberries.

Iridaceae

Iris cristata

Dwarf Crested Iris - not nearly so large as the cultivated bearded iris which is Tennessee's state flower, wild irises are no less attractive and are a pleasure to behold when encountered along stream banks in rich woods - photographed on April 6, 2011 at Carter Cave State Natural Area - Franklin County, TN. Leaves linear up to 1" wide and 6" long. Flowers 3-merous, light blue to lavender. The 3 sepals have a yellow crown adjacent to a patch of white. The petals are lacking such decoration. April - May. Fruit a capsule encasing orange-ish seeds.

Yellow Stargrass

Blackberry Lily

Dwarf Crested Iris

37

Iridaceae

Sisyrinchium albidum

Pale Blue-Eyed Grass - this common cedar glade perennial (which is not a grass at all) could easily be called yellow-eyed grass (or iris) because of the vivid yellow pollen on its anthers and the yellow throat formed at the base of the corolla - photographed on April 14, 2011 at Flat Rock Cedar Glade, Rutherford County, TN. Leaves narrowly linear. Flowers with 6 tepals, light blue (occasionally white) with a distinct yellow center. April - May.

Liliaceae

Clintonia umbellulata

Speckled Wood Lily - a perennial that prefers acidic soils, the young leaves of Speckled Wood Lily are edible and taste like cucumbers - photographed May 24, 2011 along Appalachian Trail near Rocky Bald close to Tellico Gap, Nantahala National Forest, NC. Leaves basal, 2-5 (most often 3) in number, elliptic to ovate. Flowers white, occasionally "speckled" with dark markings, with 3 sepals and 3 petals and protruding stamens. Inflorescence borne on an umbel. April - May.

Liliaceae

Erythronium albidum

White Trout Lily - this perennial is found in the woods around cedar glades and is particular spectacular since this species tends to produce relatively large stands which are in bloom more or less at the same time - photographed on March 24, 2011 at Flat Rock Cedar Glades, Rutherford, County, TN. The 2 leaves are opposite and mottled, likely giving rise to the common name trout lily. The single flower has 6 tepals which are greatly recurved at maturity. March - April. Fruit a capsule

Pale Blue-Eyed Grass

Speckled Wood Lily

White Trout Lily

Orchidaceae

Goodyera pubescens

Downy Rattlesnake Plant - The root tea of this woodland perennial was used to treat pleurisy (an inflammation of the lining of the lungs and chest) and snakebites while the leaf tea was taken with whiskey to improve appetite - photographed July 23, 2011 at Laurel Snow Pocket Wilderness, Rhea County, TN. Leaves arranged in a basal rosette and are quite beautiful, being dark green with a light stripe up the middle surrounded by a network of light veins. Flowers are hirsute, round, borne on a raceme atop a stout scape. July - August.

Orchiaceae

Spiranthes lacera var. *gracilis*

Slender Ladies' Tresses - this orchid which can grow to 16" is common in the surrounding woods and grassy areas of some cedar glades in the late summer and fall after many other noticeable wildflowers have completed their annual life cycle- photographed September 27, 2011 at Flat Rock Cedar Glades, Rutherford, County, TN. Leaves arranged in a basal rosette which withers before flowering. Flowers white with a green throat, spirally arranged around a spike-like scape in a manner suggesting braided hair which was the inspiration for "tresses" in the common name. July - October.

Poaceae

Chasmanthium latifolium

River Oats, Spangle Grass - formerly *Uniola latifolia* - *latifolium* comes from *lati-* for wide and *folium* for leaves refers to the broad leaves of this grass - this grass is hardy and establishes readily when planted along river banks and mesic woods as part of native plant restoration projects - photographed October 4, 2011 at Flat Rock Cedar Glades, Rutherford County, TN. Leaves alternate and lanceolate. Flowers are small, borne in flattened spikelets on a drooping panicle. June - October. Fruits are grains released after the spikelets have dried and opened.

Downy Rattlesnake Plant

Slender Ladies' Tresses

River Oats

41

Poaceae

Schizachyrium scoparium

Little Blue-Stem Grass - This tall (to 4') perennial grass covers hundreds of thousands of acres of prairies and is also a common sight in cedar glades where it is encountered growing between fissures in the limestone - it is widely planted by the National Park Service in an effort to reestablish native grasses in fallow fields - some horticultural varieties are used for landscaping, the plant being very drought tolerant - photographed October 4, 2011 at Flat Rock Cedar Glades, Rutherford County, TN. Leaves alternate, lanceolate to 14", slightly aquamarine in color.

Trilliaceae

Trillium undulatum

Painted trillium - the only trillium with two colors on its petals, Painted Trillium is a mountain species -photographed May 23, 2011 at Devil's Courthouse on Blue Ridge Parkway, NC. The 3 leaves are whorled and ovate. Flowers have 3 linear green sepals and 3 white petals with a wavy (undulating) edge. The base of each petal is marked by a dark pink line. April - May.

Agavaceae

Yucca filamentosa

Spanish Bayonet - this plant, reminiscent of something one would expect to find in the Southwest, is found sporadically throughout Tennessee. Often planted as an ornamental, yucca is probably a native species as evidenced by its symbiotic relationship with the yucca moth which will lay its eggs in no other host plant. In exchange the yucca is pollinated by the moth. Photographed January 24, 2014 at Flat Rock Cedar Glades, Rutherford County, TN. The stiff leaves are in a basal rosette, liner and sharply acuminate. Dozens of large, creamy white flowers borne on a tall panicle. June - September. Fruit a capsule.

Little Blue-Stem Grass

Painted Trillium

Spanish Bayonet

43

Cedar glade in early summer.

Wildflowers Illustrated

Dicots

Acanthaceae

Justicia americana

American Water Willow - this low growing plant is common along rivers throughout the eastern U.S. - provides good cover for fish and...snakes! - photographed July 4, 2011 on Duck River in Marshall County, TN. Leaves linear to lanceolate, arranged opposite on the stem. Flowers bilabiate, June - September. Fruit a 4-seeded capsule.

Acanthaceae

Ruellia caroliniensis

Hairy Petunia - a prolific perennial that makes an attractive addition to the rocky openness of summertime cedar glades - photographed July 2, 2011 at Stones River National Battlefield. Leaves are lanceolate to ovate, margins entire. Flowers funnelform with distinct petioles borne in leaf axils. May - September. Fruit a capsule.

Acanthaceae

Ruellia humilis

Glade Wild Petunia - a somewhat infrequent plant, the species name *humilis* refers to the plant's low growing habit - note the hairy, sessile leaves - photographed May 9, 2011 at Flat Rock Cedar Glade, Rutherford County, TN. Leaves similar but smaller than those of *R. caroliniensis*. Flowers are similar but smaller as well, May - July.

Water Willow

Hairy Petunia

Glade Wild Petunia

47

Apiaceae

Conium maculatum

Poison Hemlock - a tall plant identified by the purple blotches on stem - an introduced species from Eurasia, this plant was the poison that killed Socrates - medicinally used as folk cancer remedy and sedative - WARNING: contact can cause dermatitis and ingestion can be lethal - photographed June 12, 2011 at Stones River National Battlefield. 2 - 3 meters tall with hollow stems. Leaves alternate, pinnately compound. Flowers in compound umbels, May - June. Fruit a schizocarpoid capsule.

Apiaceae

Daucus carota

Queen Anne's Lace - as is clearly visible in this photo, this introduced plant is an attractive but common roadside weed - photographed on May 12, 2011 on side of Factory Rd., Rutherford Co., TN. Leaves alternate and compound; highly dissected with leaflets linear to lanceolate. Flowers arranged in an umbel, May - September. Fruit a schizocarp.

Asclepiadaceae

Asclepias incarnata

Swamp Milkweed - one common name for this plant is Indian hemp because of the fibrous nature of the stems - Native Americans used these fibers to make fishing line and thread - the roots have been added to liquor to make bitters - photographed August 7, 2011 at Flat Rock Cedar Glade, Rutherford County, TN. Leaves linear-lanceolate to vaguely ovate with distinct petioles, arranged oppositely. Flowers in umbel-like arrangements, June - August. Fruit is a dehiscent follicle releasing a seed with hairy tufts which aid in wind dispersal.

Poison Hemlock

Queen Anne's Lace

Swamp Milkweed

Asclepiadaceae

Asclepias syriaca

Common Milkweed - flower buds and shoots are edible if properly prepared - the milky sap was once chewed like gum but this practice is discouraged due to potential toxicity - Native Americans used a concoction that included this plant as a contraceptive - photographed June 12, 2011 at Stones River National Battlefield. Leaves are opposite, oblong to oval. Flowers in robust, spheric umbels, June - August. Fruit is a long, spiny follicle.

Asclepiadaceae

Asclepias tuberosa

Butterfly Weed - also called Pleurisy root, a tea of the root has been used to treat lung inflammations and other respiratory ailments - note that the milky latex is absent - photographed June 10, 2011 at Flat Rock Cedar Glades. Leaves alternate, linear-lanceolate. Flowers orange, in terminal umbels, June - August. Fruit is a follicle.

Asclepiadaceae

Asclepias verticillata

Whorled Milkweed - the linear leaves are in whorls of 3 to 6 around the narrow stem - the genus name *Asclepias* is in honor of Asclepius, the Greek god of healing - photographed August 13, 2011 at Flat Rock Cedar Glades, Rutherford County, TN. Flowers in small umbels, June - August. Fruit a narrow follicle.

Common Milkweed

Butterfly Weed

Whorled Milkweed

Asclepiadaceae

Asclepias viridis

Antelope-horn Milkweed - Leaves ovate to oblong, round or obtuse at the tip. Flowers greenish with a purple center, late April - June. Fruit a follicle. Photographed at Cedars of Lebanon State Park, Wilson County, TN on April 30, 2011.

Asteraceae

Amphiachyris dracunculoides

Prairie Broomweed - this naturalized weed from the Southeastern U.S. is spreading and has established itself in cedar glades - photographed September 10, 2011 at Flat Rock Cedar Glades, Rutherford County, TN. Leaves alternate, linear with entire margins. Flowers with sticky heads arranged in cymes, July - October. Fruit an achene.

Asteraceae

Aster divaricatus

White Wood Aster - photographed July 23, 2011 at Laurel Snow Pocket Wilderness, Rhea County, TN. Leave alternate, petioled, cordate and serrate. Ray flowers less than ten and white, disk flowers yellow, arranged in a corymb, July - October. Fruit an achene.

Antelope-horn Milkweed

Prairie Broomweed

White Wood Aster

Asteraceae

Aster paludosus var. *hemisphericus*

Southern Prairie Aster or Tennessee Aster - formerly *A. hemisphericus*, *Eurybia hemisperica* - Flowers blue to violet - Leaves alternate and linear - photographed September 27, 2011 at Flat Rock Cedar Glades, Rutherford, County, TN. Leaves alternate and linear. Ray flowers violet, disk flowers purple, borne in leaf axils, August - October.

Asteraceae

Cichorium intybus

Chicory - a familiar roadside plant - the flowers wither in the afternoon sun - used for flavoring in coffee and even cookies - the leaves can be used in salads and the flowers make a colorful garnish - medicinally used as a heart tonic and sedative - the tea promotes appetite and relieved indigestion - photographed July 9, 2011 at Stones River National Battlefield. Leaves alternate and pinnately lobed. Disk flowers absent. Ray flowers blue with toothed ends, usually closing by afternoon, June - October.

Asteraceae

Conoclinium coelestinum

Mistflower - some communities of this plant have pinkish to purple flowers - the specific epithet coelestinum is Latin for "heavenly", referring to the color of the flowers - photographed August 12, 2011 at Stones River National Battlefield. Leaves opposite, ovate deltoid to nearly cordate. Ray flowers absent. Disk flowers blue to pinkish, borne on corymbs, July - October.

Southern Prairie Aster

Chicory

Mistflower

55

Asteraceae

Coreopsis auriculata

Lobed Tickseed - The flat seeds of *Coreopsis* species are reminiscent of ticks - note the pubescence on the stem - photographed on May 9, 2011 at Flat Rock Cedar Glade, Rutherford County, TN. Leaves mostly basal and elliptic. Tips of the ray flower are 4-lobed, April - June.

Asteraceae

Coreopsis tinctoria

Garden Coreopsis - used to treat diarrhea and to induce vomiting - photographed June 12, 2011 at Stones River National Battlefield. Leaves opposite and linear. The outer two-thirds of the ray flowers are yellow but turn brick red toward the base. Disk flowers are the same. Borne on a corymb, June - September. Fruit is an achene.

Asteraceae

Echinacea simulata

Prairie Purple Coneflower - this rare species is only found in Rutherford County in TN - all members of this genus exhibit medicinal properties - rather than taken continuously *Echinacea* is most effective only when taken at the onset of a cold - photographed June 3, 2011 at Flat Rock Cedar Glades, Rutherford County, TN. Leaves hairy, thick, and mostly basal, lanceolate. Ray flowers narrow with toothed ends and reflexed. Disk flowers stiff and greenish brown, June - July. When the fruit, which is an achene is ripe, it is a favorite food of goldfinches.

Lobed Tickseed

Garden Coreopsis

Prairie Purple Coneflower

Asteraceae

Echinacea tennesseensis

Tennessee Coneflower - a highly endemic species, the cedar glades where this plant grew naturally may now sit at the bottom of Percy Priest Lake while a Rutherford County population was destroyed in 1968 to make room for a trailer park - through planting in other cedar glades the plant was saved and was recently removed from endangered status - note the upturned ray flowers, different from any other species of *Echinacea* - photographed July 2, 2011 at Stones River National Battlefield, sprouted from specimens planted by Dr. Thomas Hemmerly and others. Hairy stem with leaves similar to *E. simulata*. Flowers May - October.

Asteraceae

Erigeron strigosus

Lesser Daisy Fleabane - One will encounter this wildflower often on a walk through the cedar glades in any season. The narrow leaves make it easy to tell this common cedar glade species apart from the larger leaved and generally more robust common fleabane (*E. philadelphicus*). Photographed May 26, 2012 at Flat Rock Cedar Glades, Rutherford County, TN. Lower leaves a basal rosette which often persists through winter. Upper leaves alternate and sessile, linear or narrowly lanceolate. Ray flowers numerous and white, disk flowers yellow. May - September.

Asteraceae

Eupatorium altissimum

Tall Thoroughwort - the species name *altissimum* means "towering" - photographed September 10, 2011 at Flat Rock Cedar Glade - Rutherford County, TN. Leaves opposite, lanceolate with three palmate veins. Ray flowers absent. Disk flowers in flat topped corymbs, August - October. Fruit a resinous achene.

Tennessee Coneflower

Lesser Daisy Fleabane

Tall Thoroughwort

59

Asteraceae

Grindelia lanceolata

Lanceleaf Gumweed - the name refers to the plant's highly resinous bracts - a tea made form this aromatic plant has been used to treat asthma and coughs - photographed July 29, 2011 at Flat Rock Cedar Glades, Rutherford County, TN. Leaves alternate, lanceolate, serrate . Flowers a head, June - September.

Asteraceae

Helenium autumnale

Autumn Sneezeweed, Staggerwort - Native Americans used a snuff of the dried flower heads to treat colds. The common name Staggerwort suggests more lively uses for this plant - photographed September 27, 2011 at Flat Rock Cedar Glades, Rutherford, County, TN. Leaves alternate, lanceolate, toothed. Ray flowers reflexed and sparsely distributed, September - October. Fruit a hairy achene.

Asteraceae

Helianthus hirsutus

Stiff-haired Sunflower - several species of *Helianthus* occur in our area - the genus name combines the Greek words *helios* for sun and *anthos* for flower - all bloom during the hottest days of summer - photographed July 29, 2011 at Flat Rock Cedar Glade, Rutherford County, TN. Leaves opposite, lanceolate to ovate with short petioles and hispid. Margins vary. Flowers in head up to 3 in., July - October.

Lanceleaf Gumweed

Autumn Sneezeweed

Stiff-Haired Sunflower

Asteraceae

Leucanthemum vulgare

Oxeye Daisy - also classified as *Chrysanthemum leucanthemum* - the genus and species name for this attractive, introduced species literally means "common white flower" - photographed May 9, 2011 at Flat Rock Cedar Glade, Rutherford County, TN. Leaves mostly basal; oblanceolate (spatulate) and much lobed. Flowers May - October.

Asteraceae

Liatris cylindracea

Cylindric Blazing Star - a member of the Asteracea whose ray flowers are absent - a rather rare flower of limestone glades - photographed July 29, 2011 at Flat Rock Cedar Glade, Rutherford County, TN. Leaves alternate and linear with a pronounced mid-vein. Ray flowers absent, disk flowers purple to pink, July - September.

Asteraceae

Polymnia canadensis

White-Flowered Leafcup or Bear's Foot - "leafcup" refers to the fact that water beads up on the surface of the leaf when cupped in the hand - leaves are edible raw or cooked - photographed June 28, 2011 at Stones River National Battlefield. Leaves are opposite, broadly lobed and large. Ray flowers are white and may be many to very few. Blooms June - October.

Oxeye Daisy

Cylindric Blazing Star

White-Flowered Leafcup

63

Asteraceae

Ratibida pinnata

Grey-Headed Coneflower - an attractive ornamental, this plant become shaggy and difficult to control if left unmanaged. Crushing the head will produce an anise-like odor -photographed May 26, 2012 at Flat Rock Cedar Glades, Rutherford County, TN. Leaves are alternate, pinnately compound, and dissected. Ray flowers dramatically reflexed. Disk flowers grey to brownish on ovate head, June - August.

Asteraceae

Rudbekia hirta

Black-Eyed Susan - a popular native ornamental, the common name Black - Eyed Susan comes from an early 18th century British poem by John Gay in which Susan goes looking for her lover, Sweet William, before he sets sail - Native Americans used the root tea to treat worms - photographed July 2, 2011 at Stones River National Battlefield McFadden Farm Unit. Leaves Alternate and hirsute. Flowers July - October.

Asteraceae

Rudbeckia triloba

Thinleaf Coneflower - this common plant has leaves with three lobes though this is not always obvious - photographed July 29, 2011 at Flat Rock Cedar Glade, Rutherford County, TN. Leaves alternate and lanceolate. Flowers in heads borne on branched clusters, June - October.

Grey-Headed Coneflower

Black-Eyed Susan

Thinleaf Coneflower

Asteraceae

Senecio anonymus

Small's Ragwort - when not in flower the purple-ish undersides of the toothed basal leaves is a good identifying characteristic. Plants of the genus *Senecio* have a long history of medical use from preventing pregnancy to easing childbirth, to treating heart, lung and urinary tract ailments. This source of this plants medicinal uses are physiologically active pyrrolizidine alkaloids that can be quite toxic- photographed on May 9, 2011 at Flat Rock Cedar Glades, Rutherford County, TN. Leaves mostly basal, oblanceolate and deeply lobed. Flowers May - June. Fruit and achene with a pappus of white bristles which is the source of the genus name *Senecio* which is Latin for "old man".

Asteraceae

Solidago spp.

Goldenrod - despite being the showiest and most prominent wildflower of the fall allergy season goldenrod is not generally the culprit of sneezes and stopped up sinuses. That distinction goes to ragweed (*Ambrosia* spp.) which blooms at the same time- once used to treat uro-genital infections, *Solidago* is derived from *solido* which means "to cure", as in "make whole" hence words such as solidarity and the legal term "in solido" meaning a binding contract - several species of goldenrod occur in the central south, varying in height from 2 - 7 ft. - photographed September 10, 2011 at Flat Rock Cedar Glades - Rutherford County, TN. Leaves alternate and generally lanceolate. Flowers in a more or less plume-like inflorescence, August - October.

Asteraceae

Verbesina virginica

Frostweed, White Crownbeard - grows to 7ft. tall. Distinctive wings along the stems play a role in protecting the plant against the first heavy frosts of autumn, creating crystalline "ice flowers" near the ground - photographed August 12 2011 at Stones River National Battlefield. Leaves alternate, broadly lanceolate to elliptical, petioles alate. Flowers in bunched heads at top of stem, August - October.

Small's Ragwort

Goldenrod

Frostweed

67

Asteraceae

Vernonia gigantea

Tall Ironweed - formerly *Vernonia altissima* - grows to 10 ft. tall - a conspicuous harbinger of late summer and the coming school year - leaf tea used by Indians to ease the pain of pregnancy and childbirth - photographed August 12, 2011 at Stones River National Battlefield. Leaves alternate, lanceolate, with acuminate tips. Ray flowers absent, disk flowers in many corymbs, August - November.

Betulaceae

Podophyllum peltatum

Mayapple - often occurring in large stands this plant prized by herbalists for its many medicinal qualities and being studied by science for its anti-cancer properties. Resin from the root is used in the treatment of warts. The powerful compounds which make the plant useful to those knowledgeable of plant lore render the plant deadly toxic to unskilled hands. The fruit is edible when ripe and has a faint strawberry taste that is prized in jellies - photographed on April 21, 2009 at Radnor Lake, Davidson County, TN. Leaves are large and palmate, with one leaf on each end of a forked stem. Each plant has a single, nodding flowers with 6 - 9 white petals, April - May.

Boraginaceae

Heliotropium tenellum

Pasture or Slender Heliotrope - The genus name *Heliotropium* for this delicate annual is a combination of the Greek words *helios* (sun) and *trope* (to turn), which is what these tough plants do, living out their lives in cedar glades during the hottest months of the summer - photographed July 9, 20112 at Stones River National Battlefield. Leaves alternate and linear. Flowers small, 5-petaled, June - September. Fruit a nutlet.

Tall Ironweed

Mayapple

Slender Heliotrope

69

Boraginaceae

Lithospermum canescens

Hoary Puccoon - *Buglossoides arvensis* - this perennial of open woods and grassy areas was used by Native American to makes dyes; a root was used to reduce inflammation in the joints - photographed April 6, 2011 at Carter Cave State Natural Area - Franklin County, TN. Leaves alternate, lanceolate, with rounded tip, and covered with soft, dense hairs as the name "hoary" indicates. Flowers yellow to orange, 5-lobed, April - May. Fruit a nutlet.

Boraginaceae

Mertensia virginica

Virginia Bluebells or Cowslip - a plant of moist places, the flowers bud out pink but turn light blue as they open; though ephemeral in its season, withering soon after flowering, this plant is a perennial and an attractive addition to native gardens - photographed on April 21, 2009 at Radnor Lake, Davidson County, TN. Leaves alternate, oblanceolate to obovate. Flowers nodding with five shallow lobes, March - May. Fruit is four nutlets.

Brassicaceae

Dentaria laciniata

Cutleaf Toothwort - edible raw or cooked with a peppery taste; Hemmerly points out that the root was once chewed to alleviate toothaches - formerly classified in the genus *Cardamine* - photographed on March 24, 2011 at Flat Rock Cedar Glades, Rutherford County, TN. Leaves petioled in whorls of three, dissected, each leaflet broadly toothed. Flowers 4-petaled in terminal clusters, March - May. Fruit a silique.

Hoary Puccoon

Virginia Bluebells

Cutleaf Toothwort

71

Brassicaceae

Leavenworthia stylosa

Long-Styled Glade Cress or Nashville Mustard - a glade endemic and harbinger of spring, this plant has the largest flower of the Leavenworthias, measuring to up to 0.5 in. with an agreeable fragrance - -Leavenworthia species epitomize the life cycle of many glade endemics, sprouting and blooming early in spring, the seeds lying dormant through the heat of the summer to sprout in late fall, overwintering as a basal rosette - photographed March 8, 2011 at Flat Rock Cedar Glades, Rutherford County, TN. Leaves in a basal rosette, roundish to oblong, pinnately lobed. Flowers ranging from yellow to white with a yellow throat to lavender. Late January or February - May. Fruit a silique ~ 1.0 in., responsible for the specific epithet.

Brassicaceae

Paysonia stonensis

Stones River Bladderpod - formerly *Lesquerella stonensis* - a very rare endemic, occurring only along the Stones River and certain tributaries but only permanently established in two sites, one of which, a field by the landfill at Walter Hill in Rutherford County, Tennessee, is owned by the Tennessee Department of Environment and Conservation. The plant requires soil disturbance to germinate and seeds can lay dormant for at least six years. Other local bladderpods are the Duck River bladderpod and Short's bladderpod, both of which have yellow flowers. Photographed on April 14, 2011. Leaves mostly basal, pinnately lobed, stem leaves alternate, clasping stem. Flowers white, borne on terminal racemes, February - April. Fruits are hairy, round, air-filled pod (silicles).

Cactaceae

Opuntia humifusa

Prickly Pear - the fruit and green pads are edible when peeled and boiled. Many web pages exist detailing how to make refreshing (and occasionally tequila-infused) drinks from the sweet fruit - the spines are actually highly modified leaves with smaller spiny leaves called glochids which can be irritating to the tough - photographed June 3, 2011 at Flat Rock Cedar Glades, Rutherford County, TN. Flowers are large, attractive, yellow with a red center, May - June. Fruit is a red fleshy berry.

Long-Styled Glade Cress

Stones River Bladderpod

Prickly Pear

73

Campanulaceae

Campanula americana

Tall Bellflower - note the long style - used by Native Americans to treat tuberculosis and coughs - photographed June 28, 2011 at Stones River National Battlefield. Leaves alternate and lanceolate, and dentate. Flowers 5-lobed with a long style, borne on a terminal raceme, July - October.

Campanulaceae

Lobelia appendiculata var. *gattingeri*

Gattinger's Lobelia - photographed on May 9, 2011 at Flat Rock Cedar Glades...named for Augustine Gattinger, a physician who wrote the first flora of Tennessee, published about 1901. Leaves alternate and elliptic. Flowers 5-petaled with lower lobe 3-lipped, upper lobed 2-lipped and much smaller. May-June, but individuals often found on into a warm autumn. Fruit a capsule.

Campanulaceae

Lobelia cardinalis

Cardinal Flower - a tall perennial, to 48", the distinctive red flowers that give the plant its name are white on some specimens - Cardinal Flower was used by Native Americans to treat a variety of ailments, most notably to expel worms - highly toxic and poisonous - photographed August 12, 2011 at Stones River National Battlefield. Leaves alternate and lanceolate. Unlike other local Lobelia species which all have blue flowers the bloom of Cardinal Flower is vividly red (occasionally pink or white); two narrow upper lobes and three larger lower lobes. July - September.

Tall Bellflower

Gattinger's Lobelia

Cardinal Flower

Caprifoliaceae
Lonicera sempervirens
Trumpet Honeysuckle - a twining vine closely related to the Japanese honeysuckle of fences and waste places, the juice can be used to treat bee stings - photographed at Flat Rock Cedar Glades May 19, 2011. Leaves opposite and glaucous, perfoliate. Flowers red or yellow with yellow throat, tubular and as such are a favorite food of hummingbirds. May - July. Fruit is a red berry.

Caprifoliaceae
Sambucus canadensis
Elderberry - a tall shrub to 10', the berries used for jellies and to make wine - the berries are high in vitamin C and have been used to treat colds and flu as well as a poultice for sores and cuts - photographed June 30, 2011 at Stones River National Battlefield - Leaves Opposite and compound with highly serrate leaflets ranging from lanceolate to ovate. Flowers white atop terminal cymes 6 - 8" across. July - August. Fruit a dark purple berry.

Caryophyllaceae
Arenaria patula
Glade Sandwort - sometimes classified in the genus *Minuartia*, this plant is abundant in glades though it often goes unnoticed due to its ethereal habit, when looking at one it often seems there is more air than plant - photographed on May 9, 2011 at Flat Rock Cedar Glades, Rutherford County, TN. Leaves short and very thin. Flowers white, 5-petaled. April - June.

76

Trumpet Honeysuckle

Elderberry

Glade Sandwort

77

Caryophyllaceae

Silene stellata

Starry Campion, Widow's Frill, Thurman's Snakeroot - the latter name comes from the idea that the root could provide an antidote to snakebite due to the blotches on it which are reminiscent of rattlesnake and copperhead markings, i.e., the doctrine of signatures - a tall perennial to 4' - photographed July 23, 2011 at Laurel Snow Pocket Wilderness, Rhea County, TN. Leaves whorled, lanceolate to ovate. Flowers white with five deeply dissected petals. July - September. Fruit a capsule.

Clusiaceae

Hypericum frondosum

Golden St. Johnswort - this is a shrubby species of Hypericum with scaly, woody branches; it is commonly found along the edges of cedar glades - plants of the genus have been used to treat depression and nervous disorders - to photographed June 10, 2011 at Flat Rock Cedar Glades, Rutherford County, TN. Leaves opposite, oblong and stiff to the touch. Flowers bright yellow, 5-petaled with numerous long stamens. May - July. Fruit a capsule.

Clusiaceae

Hypericum sphaerocarpum

Round-Fruited St. Johnswort - the common name St. Johnswort is a holdover from earlier times when the Church replaced the pagan names of many plants with Christian names. Moreover, during the Dark Ages St. Johnswort was used during exorcisms to cast off evil spirits - photographed July 9, 2011 at Stones River National Battlefield. Leaves opposite, linear to slightly oblong. Flowers 5-petaled with numerous long stamens atop a corymb. June - August. As the specific epithet indicates the fruit is a round capsule.

Starry Campion

Golden St. Johnswort

Round-Fruited St. Johnswort

Crassulaceae
Sedum pulchellum

Widow's Cross, Glade Stonecrop - a low growing winter annual, these plants can be found in large stand during the spring which become dried out, crunching under foot as the heat of summer approaches - photographed May 19, 2011 at Flat Rock Cedar Glades, Rutherford County, TN. Leaves alternate, linear and succulent. Flowers 4-petaled borne on a 5-branched inflorescence that is reminiscent of a starfish. Late March - May.

Ericaceae
Kalmia latifolia

Mountain Laurel - an attractive upland species with evergreen leaves ideal for landscaping eastern and northern exposures of a house - its range extends farther west than the rhododendrons - highly toxic - photographed on May 1, 2011 at Stone Door in Grundy County, TN. Leaves oval and glaucous. Flowers cup shaped, white with hints of pink on the ribs. May - June.

Ericaceae
Rhododendron calendulaceum

Flame Azalea - though not found around the cedar glades which are the focus of this book, flame azalea is none the less an attractive plant which will be long remembered if encountered on an early summer excursion into the mountains - photographed May 24, 2011 on FS Road 1365 near its intersection with the Appalachian Trail, Nantahala National Forest, NC. Leaves deciduous and elliptic and lanate. Flowers a vivid yellow to orange as the common name indicates, reaching their peak just as the leaves are unfolding. April - June.

Glade Stonecrop

Mountain Laurel

Flame Azalea

81

Euphorbiaceae

Croton capitatus

Woolly Croton or Hogwort - eaten by game birds but poisonous to livestock - found in the hottest, driest of cedar glades - photographed August 13, 2011 at Flat Rock Cedar Glades, Rutherford County, TN. Leaves are alternate, oblong and lanate with a silvery sheen. Flowers are small, borne on a compact terminal inflorescence. June - October. Fruit is a capsule.

Euphorbiaceae

Croton monanthogynus

Prairie Tea - In his *Wildflowers of the Central South* Dr. Thomas Hemmerly describes this plant as "unspectacular" and from the picture it is obvious why one would make such an assessment. But the plant has a history of medicinal use, the leaves once having been used to make a tea though the plant is somewhat toxic - photographed July 2, 2011 at Stones River National Battlefield. Leaves similar to *C. capitatus* but smaller and not as hairy. Flowers small and white to dun colored. June - October.

Euphorbiaceae

Euphorbia corollata

Flowering Spurge - a relatively tall perennial (to 40") like other members of the Euphorbiaceae this plant exudes a milky sap when a stem is broken; this sap has been known to cause dermatitis - A tea made from the root is a strong laxative and emetic, which gives rise to an archaic common name for the plant: Purging Root - photographed July 29, 2011 at Flat Rock Cedar Glades, Rutherford County, TN. Leaves are alternate a narrowly elliptic, whorled at the branches. Flowers are small and inconspicuous but are surrounded by 5 white, petal-like bracts born atop an open cyme. July - September.

Woolly Croton

Prairie Tea

Flowering Spurge

Fabaceae

Astragalus bibullatus

Pyne's Ground Plum - a rare cedar glade endemic which has only been found in Rutherford, County Tennessee - not described until 1987 when Milo Pyne decided it was likely a distinct species - much work has been done by the Missouri Botanical Gardens to try to preserve this rare plant which competes poorly with surrounding plants and is subject to deer grazing - when not in flower it can be differentiated from Tennessee Milk Vetch (*A. tennesseensis*) by its glabrous stems and leaves, those of Tennessee Milk Vetch being pubescent - photographed March 18, 2012 at Flat Rock Cedar Glades, Rutherford Co., TN. Leaves Alternate and pinnately compound with elliptic leaflets. Flowers are a showy lavender. March or April depending on the whether or not it is a warm spring. Fruit is a smooth pod that superficially resembles a plum.

Fabaceae

Astragalus tennesseensis

Tennessee Milk Vetch - a relatively infrequent member of the Fabaceae, Astragalus species are noted for their anti-viral properties. Huang-qi, an Asian species has been used in Chinese medicine for thousands of years. One author notes that these plants can accumulate selenium in their leaves causing a sickness in livestock called loco disease. Note the hirsute stems and leaves which set it apart from the extremely rare Pyne's Ground Plum. Photographed April 21, 2011 at Flat Rock Cedar Glades, Rutherford County, TN. Leaves alternate, pinnately compound and elliptic. Flowers a creamy yellow. April - May. Fruit a curved legume.

Fabaceae

Baptisia australis

Blue False Indigo - A tall (to 5'), herbacious perennial whose flowers were once used to create a blue dye- the species name australis means "southern" - infrequent - photographed May 10, 2011 at Flat Rock Cedar Glades Rutherford County, TN. Leaves alternate and trifoliate, i.e. similar to a clover; leaflets obovate. Flowers are pea-like (papilionaceous), large and deep blue. May - June. Fruit is a plump legume with seeds that rattle loudly once it dries out.

84

Pyne's Ground Plum

Tennessee Milk Vetch

Blue False Indigo

85

Fabaceae

Dalea gattingeri

Gattinger's Prairie Clover - a low growing perennial with wispy, thin leaves that are highly aromatic - this plant can be the dominant plant cover in gravelly glades during the driest months - photographed at Flat Rock Cedar Glades May 19, 2011. Leave alternate and pinnately compound with linear leaflets. Flowers are small and purple developing on a long cylindrical head. July - September.

Fabaceae

Desmanthus illinoensis

Prairie Mimosa - A tall perennial which can grow to 5', the fruit is a curious globular bundle of little pods, hence the genus name which is from the Greek *desme* for "bundle" - fresh roots were chewed for toothaches by Native Americans while a leaf poultice was used for ear aches - photographed June 25, 2011 at Flat Rock Cedar Glades, Rutherford, County, TN. Leaves alternate and bipinnately compound, suggestive of a mimosa tree (*Albizia julibrissin*). Flowers are small and white arranged in a ball-like inflorescence with protruding stamens. June - July.

Fabaceae

Desmodium nudiflorum

Naked Flowered Tick Trefoil, Beggars Lice - the Cherokee chewed the root to relieve sore gums - photographed July 23, 2011 at Laurel Snow Pocket Wilderness, Rhea County, TN. Leaves sparse, trifoliate and ovate. Flowers are pea-like on a racemose inflorescence. July - August. Seeds are within a flattened legume that breaks into individual segments at maturity; short hooked hairs cause them to stick to fur and clothing.

Gattinger's Prairie Clover

Prairie Mimosa

Naked Flowered Tick Trefoil

Fabaceae

Lathyrus latifolius

Everlasting Pea - an escaped garden perennial introduced from Europe - planted by Thomas Jefferson at Monticello in 1807 - photographed June 30, 2011 at Stones River National Battlefield. Leaves alternate and pinnately compound with winged petioles. Flowers are showy, fuchsia to white. June - August. Fruit is a smooth pod to 4" long.

Fabaceae

Pediomelum subacaule

Nashville Breadroot - a showy perennial that is quite striking when encountered in cedar glades in the spring - the tuber of a related species was once roasted and eaten which may explain the name Nashville Breadroot - photographed on April 14, 2011 at Flat Rock Cedar Glades, Rutherford County, TN. Leaves are palmately compound with leaflets obovate to elliptic. Flowers are racemose and dark blue to purple, though occasionally white specimens may be encountered. Blooms do not appear before the fourth or fifth year. April - May. Fruit is a legume.

Fabaceae

Senna marilandica

Southern Wild Senna - formerly *Cassia marilandica* - this perennial can grow to 7 ft tall - leaves fold up in the evening - medicinally the leaf tea is a strong laxative, being a native alternative to commercially marketed Alexandrian Senna (*S. alexandrina*) - photographed August 4, 2011 at Stones River National Battlefield. Leaves are alternate and compound, elliptic. Flowers are racemose with long petals. July - August. Fruit is a legume.

Everlasting Pea

Nashville Breadroot

Southern Wild Senna

89

Geraniaceae
Geranium maculatum

Wild Geranium - this true geranium is an attractive native plant of rich woodlands; despite sharing the name the summer annual encountered in flower pots is not related - photographed May 24, 2011 along Appalachian Trail near Rocky Bald close to Tellico Gap, Nantahala National Forest, NC. Leaves are moderately scabrous with 3 to 5 acute lobes, the tips of each being toothed or deeply notched. Flowers are pink to dark rose with 5 petals and 10 stamens. April - June. Fruit a beaked capsule.

Hydrangeaceae
Hydrangea arborescens

Wild Hydrangea - has been cultivated into a popular ornamental of shady landscapes, these plants are equally attractive when encountered in rich woods growing near waterfalls or other shady ledges - photographed July 23, 2011 at Laurel Snow Pocket Wilderness, Rhea County, TN. Leaves are opposite, ovate, and finely toothed. In this species the leaves are glabrous while in other similar species they may be pubescent. Two types of flowers exist on the flat-topped inflorescence. The fertile flowers are exceedingly small. They are sparsely surrounded by a few larger, 3 to 4-petaled sterile flowers around the outer edge.

Lamiaceae
Clinopodium glabellum

Glade Savory - formerly in genus *Calamintha* and *Satureja* - a common mint of the cedar glades - used to flavor teas and to soothe upset stomachs and frayed nerves - photographed June 25, 2011 at Flat Rock Cedar Glades, Rutherford, County, TN. Leaves opposite and linear, decreasing in size further up the stem. Flowers white with a hint of pink, aromatic, growing in leaf axils. May-August.

Wild Geranium

Wild Hydrangea

Glade Savory

Lamiaceae

Isanthus brachiatus

False Pennyroyal or Fluxweed - formerly *Trichostema brachiatum* - a common wildflower of late summer in dry cedar glades - photographed September 2011 at Flat Rock Cedar Glades, Rutherford County, TN. Leaves opposite, elliptic, accuminate. The small lavender flowers are atypical of the Lamiaceae in that they are symmetrical with a 5-lobed corolla with 4 stamens. July - September.

Lamiaceae

Perilla frutescens

Beefsteak Plant - this somewhat tall (to 3') highly aromatic annual herb has a liquorice-like odor and can be used as a pot herb, especially to flavor meats, as the common name implies - medicinally the tea has been used by pregnant women to relieve morning sickness and to "calm a restless fetus" but use during pregnancy is discouraged - rubbing the leaf oil on a wart will cause the wart to disappear within days though the oil will discolor the surrounding skin for a few hours - photographed July 9, 2011 at Stones River National Battlefield. Leaves opposite and ovate with a coarse texture; note the dark burgundy to purple color. Flowers small with a 5-lobed corolla borne on a raceme, July - September. Fruit a nutlet.

Lamiaceae

Physostegia virginiana

False Dragonhead or Obedient Plant - the "obedient" in the name refers to the stiff cymes which stay in place when bent for floral arrangements; a handful of cultivars have been bred for use in home gardening - photographed September 10, 2011 at Flat Rock Cedar Glades, Rutherford County, TN. Leaves opposite, lanceolate, and serrate. Flowers large and tube shaped, 5-lobed with 2 lips above and 3-lipped below, arranged on a raceme; light pink, becoming dark pink above. July - September. Fruit an angled nutlet.

False Pennyroyal

Beefsteak Plant

Obedient Plant

93

Lamiaceae

Prunella vulgaris

Selfheal, Heal All - as the name implies this plant has a litany of medical uses - leaf tea can be gargled for sore throats - leaves can be poulticed on wounds and bruises - contains antibiotic and anti-tumor properties - photographed July 23, 2011 at Laurel Snow Pocket Wilderness, Rhea County, TN. Leaves opposite, lower leaves elliptic and petioled, upper leaves lanceolate. Flowers bilabiate arising from rough bracts arranged on a stubby cylindrical spike. May - September. Fruit a nutlet.

Lamiaceae

Scutellaria incana

Downy Skullcap - plants of this genus have been used in folk medicine to relieve "nervous" conditions and science has found that plants such as Downy Skullcap contain the anti-spasmodic scutellaine - photographed on May 9, 2011 at Flat Rock Cedar Glades, Rutherford County, TN. Leaves opposite and ovate; crenate and covered with short hairs as the descriptive name downy implies. The blue flowers are bilabiate, growing on a terminal raceme or from the upper leaf axils. May - June.

Loganiaceae

Mitreola petiolata

Mitrewort or Lax Hornpod - formerly *Cynoctonum mitreola* - this plant is cited as a wetland delineation species, earning a FACW+ as a Facultative Wetland indicator in the Southeast and Northeast, meaning that 67% or more of all occurrences will be in a wetland environment - photographed on October 20, 2012 at Flat Rock Cedar Glades. Leaves opposite, 4-ranked, and ovate and slightly scabrous. Upper leaves sessile. Flowers very small, white and 5-lobed on cymes, July - October. Fruit a capsule.

Selfheal

Downy Skullcap

Mitrewort

Loganiaceae

Spigelia marilandica

Indian Pink or Wormgrass due to a root tea of this plant being used as a worm expellant by Native Americans and frontier physicians. Use is not recommended due to serious side effects such as convulsions, increased heart rate and potentially death - photographed June 26, 2010 at Virgin Falls Pocket Wilderness, White County, TN. Leaves opposite, ovate to broadly lanceolate, lacking petioles. Flowers salverform, strikingly scarlet outside flaring at the top into 5 acuminate lobes, exposing a greenish yellow interior. May - July. Fruit is a swollen capsule that catapults seeds via hydrostatic pressure.

Malvaceae

Hibiscus moscheutos

Swamp Rose Mallow - closely related to cotton and okra, relatives of this wetlands loving plant are cultivated as ornamentals - medicinally this highly mucilaginous plant has been used to provide relief for discomfort of the urinary and gastrointestinal tracts - photographed July 15, 2011 at Stones River National Battlefield. Leaves alternate and ovate to nearly cordate, often with red veins, occasionally with shallow lobes. Flowers showy, white to faintly pink with a blood red center. June - September. Fruit is a beaked capsule.

Onagraceae

Oenothera macrocarpa

Missouri Evening Primrose - (formerly *O. missouriensis*) Considered a disjunct in our area, this plant is common in the Missouri, Oklahoma, and Texas. Though this is a rare species in our area, found only in Rutherford and Wilson counties, it is quite conspicuous when present in cedar glades due to its large, showy flowers - Missouri Evening Primrose is pollinated by the sphinx moth and generally opens in late afternoon and like other flowers is most fragrant only when open for pollination - photographed on May 9, 2011 at Flat Rock Cedar Glades, Rutherford County, TN. Leaves alternate and lanceolate on low growing stems. Flowers large, up to 4" across, 4-petaled. May - July. Fruit a capsule encased in a long winged pod which is reminiscent of a lantern.

Indian Pink

Swamp Rose Mallow

Missouri Evening Primrose

Onagraceae

Oenothera speciosa

Showy Evening Primrose - an introduced species from the mid-west, this plant is both showy and tough, resistant to drought and thriving in hot, exposed areas - photographed on May 12, 2011 on side of Factory Rd., Rutherford County, TN. Leaves alternate and lanceolate to linear; irregularly toothed. Flowers faintly pink (occasionally white) with 4 deeply veined petals. March - August.

Orobanchaceae

Conopholis americana

Squaw Root - this edible plant lacks chlorophyll and is a parasite found in association with oaks in rich forests; the alternate name Cancer Root implies that it was probably used medicinally at one point - photographed May 24, 2011 on Appalachian Train near Tellico Gap, Nantahala National Forest, NC. Leaves are fleshy and imbricated. The flowers are small, bilabiate, and white, arising from the leaf axils. April - June.

Orobanchaceae

Veronicastrum virginicum

Culver's Root - In the cedar glade country of Tennessee's Central Basin this plant of barrens and roadsides has only been found in Rutherford County though it is common in more upland areas throughout the eastern United States - the dried root used by Native Americans as a laxative and diuretic - probably toxic -photographed May 26, 2012 at Flat Rock Cedar Glades, Rutherford County, TN. Leaves whorled and lanceolate. Flowers small and white on a tapering spike. Note the 2 protruding stamens. May - September.

Showy Evening Primrose

Squaw Root

Culver's Root

Oxalidaceae

Oxalis priceae

Price's Wood Sorrel - a frequently encountered cedar glade perennial with showy flowers, note the shamrock-like leaves that have a citrusy taste when chewed - photographed May 9, 2011 at Flat Rock Cedar Glade, Rutherford County, TN. Leaves trifoliate. Flowers 1/2" or larger, yellow with a red center. April - May.

Oxalidaceae

Oxalis violacea

Violet Wood Sorrel - though not common this attractive plant is widespread and may be encountered on the fringes of cedar glades - photographed April 9, 2011 at Carter Cave State Natural Area, Franklin County, TN. Leaves reminiscent of a shamrock with a purple hue. Flowers are delicate, violet to white with 5 petals. Late March - June. Fruit a capsule.

Passifloraceae

Passiflora incarnata

Passionflower or Maypop is a long climbing vine that may cover the ground hot, open areas in late summer - the kiwi-size fruit is a berry and are edible and said to be delicious - medicinally used as an anti-spasmodic and to tame restless nerves. Passionflower is the state wildflower of Tennessee - photographed August 23, 2012 on Crews Hollow Road, Coffee County, TN . Leaves alternate and 3-lobed on a long petiole. Flowers large and showy with striking purple and white banded threads arising from 5 sepals and 5 petals. June - September.

Price's Wood Sorrel

Violet Wood Sorrel

Passionflower

Phytolaccaceae

Phytolacca americana

Pokeweed or Poke Sallet - a tall perennial herb, rarely has such a poisonous plant been so highly valued as a food source (almonds come to mind) - tender young leaves and the tips of older ones can be boiled through two changes of water then fried up with eggs or butter - Gainesboro, TN hosts an the annual poke sallet festival every May - the berries produce a dark, wine colored juice which can be used as ink - photographed June 28, 2011 at Stones River National Battlefield. Leaves alternate and elliptic, lanceolate higher up the stem. Flowers small with 5 sepals and numerous (up to 30) borne on an indeterminate raceme. May - September. Fruit a deep purple berry.

Polemoniaceae

Phlox amoena

Hairy Phlox - known as Sweet William to gardeners, phlox is an attractive wildflower of open woods and fields - photographed May 9, 2011 at Flat Rock Cedar Glades. Leaves opposite and lanceolate. Flowers 5-lobed, fuchsia to faintly lavender the calyx is hairy but the corolla tube is glabrous. April - June. Fruit a capsule.

Polygonaceae

Polygonum virginianum

Jumpseed, Virginia Knotweed - formerly *Tovara virginiana* - the name Knotweed for this sturdy perennial may come from the enlarged leaf node - the tender young leaves are edible, the flavor blending particularly well with sesame oil and ginger - photographed July 9, 2011 at Stones River National Battlefield. Leaves alternate and ovate; the petiole forms a sheath called an orcea. Flowers are white with a green tinge with 4 tepals, borne on a tall (to 24") raceme. July - October. Fruit an achene. The name Jumpseed comes from the seed dispersal mechanism which is triggered by the style.

Pokeweed

Hairy Phlox

Virginia Knotweed

103

Portulacaceae

Talinum calcaricum

Limestone Fameflower - first described as a distinct species in 1967 by Stewart Ware, the Limestone Fameflower is small but attractive perennial with blooms that open in the hottest part of the day then rapidly wilt - prefers dry, stony soils where it stores water in its thick succulent leaves - photographed June 3, 2011 at Flat Rock Cedar Glades, Rutherford County, TN. Leaves linear and succulent. Flowers 5-petaled with numerous stamens, salmon to rose in color. May - September. Fruit a capsule.

Primulaceae

Dodecatheon meadia

Shooting Star - the curious genus name of this eye-catching perennial means "12 gods" in reference to the twelve Olympians who, led by Zeus, rule from Mount Olympus; the specific epithet *meadia* is in honor of Richard Mead, an English physician - other common names include Pride of Ohio, Indian Chief, and Rooster Heads - photographed April 21, 2011 at Flat Rock Cedar Glade, Rutherford County, TN. Leaves basal,oblanceolate and thickened, leading one 18th century English gardener to think the plant was "Coss lettice". Flowers usually white but occasionally pink with 5-highly reflexed petals. April - June. Fruit a capsule with red seeds.

Ranunculaceae

Anemone virginiana

Thimbleweed - numerous medicinal uses including the treatment of tuberculosis and diarrhea. According to medicinal botanist James Duke the root was placed under the pillow of a "crooked wife" to divine the truth about what she has been doing; it may also be a crucial ingredient in some love potions - photographed June 25, 2011 at Flat Rock Cedar Glade, Rutherford County, TN. Leaves whorled, palmate with the outer edges deeply lobed. Flowers white to greenish-white, with 5 showy sepals but lacking petals, stamens numerous. May - July. Fruit borne on a large spherical compound pistil reminiscent of a thimble.

Limestone Fameflower

Shooting Star

Thimbleweed

105

Ranunculaceae

Consolida ajacis

Rocket Larkspur - a beautiful exotic found in waste places, escaped from cultivation - photographed July 15, 2011 at Stones River National Battlefield. Leaves alternate, linear to the point of being needle-like, and severely dissected. Flowers dark blue to light pink with a long spur. June - September.

Ranunculaceae

Delphinium virescens

Prairie Larkspur - the name "larkspur" comes from the flower's resemblance to the back claw on a bird's foot - this tall growing perennial comes from a genus whose other members produce seeds and saps that possess effective insecticidal and parasitidal properties - photographed on May 9, 2011 at Flat Rock Cedar Glades, Rutherford County, TN. Leaves alternate and linear. Flowers white with pink tips to faintly blue, borne in great quantities on a tall (to 4') spike. May - July. Fruit a beaked follicle.

Ranunculaceae

Thalictrum thalictroides

Rue Anemone - a delicate flower of deep woods, occurring along the woodland edge of cedar glades - The root is considered edible though it contains toxic compounds while other parts of the plant were once used to treat gastrointestinal illness - the Greek suffix *-oides* means "resembling" photographed March 18, 2012 at Flat Rock Cedar Glades, Rutherford Co., Tennessee. Leaves whorled and compound with 3 shallowly lobed leaflets. Flowers with 5-10 sepals white to pinkish in color; petals absent. March - May. Fruit an achene.

Rocket Larkspur

Prairie Larkspur

Rue Anemone

Rosaceae

Agrimonia pubescens

Soft Agrimony - the tea of some members of this genus is considered flavorful and drunk for pleasure as well as to relieve nausea - it is considered a tonic for gout - photographed July 9, 2011 at Stones River National Battlefield. Leaves stipulate, pinnately compound, leaflets elliptic to slightly oblanceolate, scabrous above, pubescent beneath. Flowers yellow, 5 petals with 10 stamens. June - August.

Rosaceae

Fragaria virginiana

Wild Strawberry - edible; high in many vitamins and minerals - photographed April 11, 2011 at Flat Rock Cedar Glades, Rutherford County, TN. Leaves low to the ground, palmate and compound with 3 toothed leaflets. Flowers white with 5 petals and numerous stamens. April - May. Fruit an achene, many of which are collected into the familiar strawberry-shape which is actually the mature receptacle of the flower.

Rosaceae

Geum canadense

White Avens - the root has purported medicinal values - boiled root added to homebrew beer is said to provide a pleasing flavor - photographed June 12, 2011 at Stones River National Battlefield. Leaves alternate, lower leaves compound and ovate with 3 leaflets. Flowers with 5 widely spaced petals and 5 conspicuous sepals. May - July. Fruit and aggregation of achenes.

Soft Agrimony

Wild Strawberry

White Avens

Rosaceae

Porteranthus trifoliatus

Bowman's Root, False Ipecac - very similar in appearance to American Ipecac (*P. stipulatus*) - despite being called False Ipecac this perennial of mesic woodlands too has medicinal value as an emetic - photographed on side of FS Road 423 at Tatham Gap, NC - May 24, 2011. Leaves alternate and compound with 3 highly toothed narrowly lanceolate leaflets. Flowers white with 5 petals. May - June. Fruit a capsule.

Rosaceae

Potentilla simplex

Common Cinquefoil - medicinally this perennial is considered an astringent and antiseptic - In 1785 Manasseh Cutler wrote "A decoction of it is used as a gargle for loose teeth and spungy gums." - photographed May 9, 2011 at Flat Rock Cedar Glades, Rutherford County, TN. Leaves alternate and palmately compound with 5 obovate leaflets, hence the common name Cinquefoil. Flowers yellow with 5 petals. April - June. Fruit an aggregate of achenes.

Rubiaceae

Houstonia serpyllifolia

Mountain Bluet - formerly *Hedyotis michauxii* - the specific epithet *serpyllifolia* of this delicate wildflower means "thyme leaved" - photographed May 24, 2011 on Appalachian Trail near Rocky Bald near Tellico Gap, Nantahala National Forest, NC. Leaves roundly ovate with short petioles. Flowers light blue to violet with a yellow center. April - June. Fruit a flattened capsule.

Bowman's Root

Common Cinquefoil

Mountain Bluet

111

Rubiaceae

Houstonia purpurea var. *calycosa*

Venus' Pride - this perennial is a common wildflower in the dry woods around cedar glades -Dr. William Houston was a Scottish botanist who, embarked as ship's surgeon, collected plants in Central America and the West Indies - photographed April 30, 2011 at Cedars of Lebanon State Park, Wilson County, TN. Leaves opposite and linear, lacking petioles. Flowers funniform, white to lavender. April - July. Fruit a round capsule.

Rubiaceae

Mitchella repens

Partridgeberry - the edible berries of this low growing perennial herb have been used as a flavoring for milk - a tea made from the leaves is said to be a good overall tonic for feeling better - photographed May 24, 2011 on Appalachian Trail near Rocky Bald near Tellico. Leaves evergreen, opposite, roundly ovate with leathery texture. Flowers funniform and fragrant, occurring in pairs with a shared ovary that produces a single red berry. May - June.

Saxifragaceae

Tiarella cordifolia

Foamflower - a popular ornamental perennial - also used to treat burns or made into a tea to treat mouth sores - photographed May 24, 2011 along Appalachian Trail near Rocky Bald close to Tellico Gap, Nantahala National Forest, NC. Leaves basal and cordate (hence *cordifolia*). Flowers 5-petaled with 10 long stamens; borne on a terminal raceme. April - June.

Venus' Pride

Partridgeberry

Foamflower

113

Scrophulariaceae

Agalinis tenuifolia

Slender Gerardia - these delicate, ephemeral flowers are a common sight in limestone glades during late summer - photographed August 13, 2011 at Flat Rock Cedar Glades, Rutherford, County, TN. Leaves opposite and linear, often dark green to nearly maroon. Flowers faint pink to light fuchsia, with a 5-lobed corolla. July - October. Fruit a capsule.

Scrophulariaceae

Aureolia patula

Spreading False Foxglove - this perennial herb is most often found along river bluffs in richly wooded areas - photographed July 23, 2011 at Laurel Snow Pocket Wilderness, Rhea County, TN. Leaves opposite, elliptic to lanceolate with winged petioles. Flowers funnelform, yellow to pale yellow with 5 lobes. July - September. Fruit a capsule with winged seeds.

Scrophulariaceae

Dasistoma macrophylla

Mullein Foxglove - *Seymeria macrophylla* - this partially parasitic perennial is the only member of its genus - photographed June 25, 2011 at Flat Rock Cedar Glades, Rutherford County, TN. Leaves opposite, ovate, and accuminate; upper leaves becoming lanceolate. Flowers 5-lobed, corolla smooth on the outside but wooly within. June - September. Fruit a beaked capsule.

Slender Gerardia

Spreading False Foxglove

Mullein Foxglove

115

Scrophulariaceae

Pedicularis canadensis

Wood Betony, Lousewort - the name Lousewort may cause one to think this plant might be used medicinally for the treatment of lice but the name actually derives from the belief that livestock grazing on the European cousin of this plant would soon develop lice - this particular species was eaten by Native Americans and early settlers, who boiled the leaves in a soup - photographed March 30, 2012 at Leatherwood Ford in Big South Fork National River and Recreation Area, Scott County, TN. Leaves basal and alternate, lanceolate with deep, narrow lobes. Flowers bilabiate, yellow to maroon, sometimes both on one flower. April - June.

Scrophulariaceae

Penstemon tenuiflorus

Slender Whiteflower Beardstongue - an attractive glade perennial, the genus name *penstemon* refers to the plant's 5 stamens - photographed at Cedars of Lebanon State Park, Wilson County, TN on April 30, 2011. The leaves basal and lanceolate, finely pubescent, entire to finely toothed. Flowers white to off-white. April - June. Hairy Beardstongue (*P. hirsutus*) is another member of this genus occasionally found growing around cedar glades. It can be distinguished from the former species by its pale violet flowers and serrated leaves.

Scrophulariaceae

Verbascum thapsus

Common Mullein, Flannel Plant, Quaker Rouge - this common exotic biennial has a long litany of utilitarian purposes - stems dipped in tallow serve as candles - hikers can use leaves to add cushion to worn out boots or provide soft toilet paper in the woods - medicinally the leaves have been smoked to relieve asthma - flowers soaked in olive oil can be used for earache drops - the seeds contain a poison that has been used to kill fish - photographed July 9, 2011 at Stones River National Battlefield. Leaves basal first year extending into second. Stem leaves alternate, elliptic to ovate; all leaves richly pubescent giving the plant a hoary, downy appearance. Flowers yellow with 5 lobes; borne at irregular intervals on a thick spike. June - September.

Wood Betony

Slender Whiteflower
Beardstongue

Common Mullein

Solanaceae

Datura stramonium

Jimsonweed - this striking, tall growing (to 5') annual is occasionally found on abandoned farm locations - members of the *Datura* genus have been used by shaman to induce terrifying hallucinogenic experiences - Jimsonweed likewise has the ability to send willing or accidental partakers into a state some have described as madness - my great-grandfather smoked Jimsonweed for asthma and this practice goes back to Ayurvedic medicine from India - photographed September 2, 2010 at Stones River National Battlefield. Leaves alternate and lanceolate to 2" wide with shallowly lobed teeth. Flowers funnelform and helical. White with a maroon center. Fruit a spiny capsule filled with seeds. Do Not Experiment With This Plant!...Leave This One Alone!

Solanaceae

Solanum carolinense

Horse Nettle, Thorn Apple - berries fried in grease can be used as an external treatment for mange - WARNING: like many members of the Solanaceae this plant is extremely toxic even though it is closely related to the potato - photographed June 12, 2011 at Stones River National Battlefield. Leaves alternate and ovate with shallow lobes, covered in potentially irritating trichomes (bristles). Flowers white to pale lavender, with 5 lobes. May - September. Fruit a yellow berry resembling a small tomato. Do not eat this!

Hyrdophyllaceae

Phacelia bipinnatifida

Purple Phacelia, Scorpionweed - a delicately flowered plant found in rich bottom lands and moist woods. One species of Scorpionweed (*P. dubia* var. *interior*) is a cedar glade endemic, found only in a handful of counties in middle Tennessee's Central Basin. It has lighter flowers than Purple Phacelia. Photographed April 21, 2009 at Radnor Lake, Davidson County, TN. The leaves are bipinnately compound, deeply cut, and borne on long petioles. The flowers are 5-lobed, being blue to violet on the outer parts of the lobe, becoming white toward the center. Stamens extend past the rim of the petals.

Jimsonweed

Horse Nettle

Purple Phacelia

119

Verbenaceae

Verbena canadensis

Rose Vervain - sometimes classified as *Glandularia canadensis* - a somewhat uncommon wildflower found along the margins of cedar glades and other dry, rocky areas - members of this genus were once used medically by both Native Americans and 19th century physicians to treat a variety of illness, including jaundice - photographed April 12, 2013 at Flat Rock Cedar Glades, Rutherford County, TN. Stems and leave pilose to hispid. Leaves ovate and deeply dissected into acute lobes. Flowers 5-lobed, lavender to pink borne in a ball-like cluster. March - September.

Verbenaceae

Verbena simplex

Narrowleaf Vervain - though this perennial is found in cedar glades, roadsides, and waste places, the genus name *Verbena* harkens back to ancient Roman, verbenae being the boughs of laurels and myrtles used in ceremonies for the gods - photographed June 12, 2011 at Stones River National Battlefield. Leaves opposite and oblong, linear farther up the stem. Flowers small, pale violet to white, 5-lobed, borne on a thin spike. May - September.

Violaceae

Viola rostrata

Long Spurred Violet - this distinctive perennial can be in rich woodlands along rivers, creeks, and other moist locations - the specific epithet *rostrata* means "beaked" and likely refers to the long spur which makes this violet readily identifiable - Photographed March 30, 2012 at Leatherwood Ford in Big South Fork National River and Recreation Area, Scott County, TN. Leaves ovate and toothed with a notch at the bottom typical of violets. Flowers 5-petaled, with two above, three below. The spur extends from the middle lower petal. April - June. Fruit a capsule.

Rose Vervain

Narrowleaf Vervain

Long Spurred Violet

Non-Flowering Plants

Cyanobacteria, Lichens, Mosses, Liverworts, Ferns, and Gymnosperms

Non-Flowering Plants

All of the plants described in the last section are flowering plants, meaning that they produce a bloom, no matter how inconspicuous it may be, such as in the grasses, which contains all the plant anatomy necessary to produce and fertilize an egg or megaspore*. Yet some of the most commonly encountered plants in cedar glades are non-flowering, meaning that they do not produce a bloom. Instead non-flowering plants produce spores which take on the roll of pollen and the egg. Some of these spores are identical, such as in the mosses and liverworts. These plants are said to be homosporous. Other plants such as Equisetum and gymnosperms produce microspores (male) and megaspores (female). These plants are called heterosporous. Furthermore all flowering plants are vascular, meaning they have specialized cells for conducting the flow of water and nutrients throughout the plant. On the other hand primitive plants such as mosses and liverworts are non-vascular, instead relying on external water to spread gametes and nutrients. This is one reason why liverworts in particular are found in association with waterfalls and seepage areas.

Other cedar glade plants are not truly plants at all but are considered so in this manual due to long standing convention. These include the cyanobacteria, algae, and lichens. These organisms are responsible for making molecular oxygen available for biological pathways. Even more importantly cyanobacteria and algae are two of the groups of organisms collectively called phytoplankton which produce at least half of Earth's oxygen. In the next few pages we'll explore the non-flowering plants most commonly encountered in glades.

*Not all flowering plants produce flowers capable of both pollinating and being pollinated. Some species of plants have individuals exhibiting pollen producing or staminate flowers while others develop flowers in which an ovary matures. These are called pistilate flowers. Plants that exhibit such a distinction between flower types are said to be dioecious. On the other hand most flowering plants are monoecious meaning that one bloom produces both fertile stamens and ovaries.

Nostoc commune

Witch's Butter, Star Jelly - *Nostoc* is a cyanobacteria and is encountered in two form in cedar glades. During the dryness of summer it is brittle, flaky and nearly black, like small pieces of charred paper. During the wet season or immediately after a rain *Nostoc* becomes an unattractive, dull green gelatinous lump. Under a microscope the true structure of *Nostoc* reveals itself as a chain of spherical cells. *Nostoc* contributes greatly to the production of biologically available nitrogen in cedar glade soils. It has attracted interest as a dietary supplement and is used as a food in Peru but some research has shown *Nostoc* to contain an amino acid linked with neurodegenerative diseases. Photographed October 3, 2009 at Flat Rock Cedar Glades, Rutherford County, TN.

Cladonia spp.

Reindeer Lichen - this interesting pale blue-grey feature of cedar glades, found in all seasons is actually the symbiotic relationship between a green algae or cyanobacteria and a fungus. The algae or cyanobacteria is called the photobiont meaning it produces nutrients using the suns energy. The fungus in the relationship is known as a mycobiont and encases the photobiont in a protective crust which protects it from the temperature and moisture extremes of cedar glades' harsh environment. Common around the world reindeer lichen has been used medicinally in Nordic cultures for the treatment of diarrhea and as a component in aquavit.

Pleurochaete squarrosa

Glade Moss - This non-vascular plant can be found on the fringes of cedar glades as well as in the heart of the glades when soil is deep enough. When encountered in all its glory on a winter day under the angled light of the afternoon sun a bed of glade moss can be quite striking for its fluorescent green color. The genus name comes from the Greek *Pleuron* (rib) and *chaete* (hair) and refers to the fact that the sporophytes are borne on the side of the vegetative structure as opposed to the top as in most mosses. Research has shown Glade Moss to possess anti-bacterial properties (Basile et al, 1998). Photographed December 9, 2011 at Flar Rock Cedar Glades, Rutherford County, TN.

Witch's Butter

Reindeer Lichen

Glade Moss

127

Cheilanthes lanosa

Hairy Lipfern - The term lipfern refers to the tendency of the edges of the leaflets to reflex around the sori borne on the underside of each leaflet. This would provide protection for developing spores in the dry conditions in which this fern is found. The specimen shown here was found growing in the shade of a small winged elm and glade privet near the edge of a spring. Photographed February 3, 2013 at Flat Rock Cedar Glades, Rutherford County, TN.

Ophioglossum engelmannii

Limestone Adder's Tongue - this fern occurs in cedar glades in spring, occasionally emerging again in late summer after a period of wet weather. With 1260 chromosomes adder's tongues have the highest chromosome count of any organism studied to date. To put it in perspective humans have 46. Photographed Flat Rock Cedar Glades, Rutherford County, TN, April 12, 2013.

Juniperus virginiana

Eastern Redcedar - this ubiquitous gymnosperm of cedar glades can be found growing as a magnificent specimen in pastures where it may attain a height of over 100 feet. More often it is encountered as a small tree growing in pure stands or as a shrub growing in a glade. Harlow and Harrar place the maximum age at around 850 years.Economically cedars are used for making pencils, furniture, and as paneling for closets due to the appealing color and pleasant odor of the tree's aromatic heartwood. Cedars are associated with cedar-apple rust, a plant disease in which fungal galls on cedar trees produce spores that are carried by wind to apple blossoms and leaves, severely damaging the yield. Photographed February 20, 2012 at Flat Rock Cedar Glades, Rutherford County, TN.

Hairy Lipfern

Limestone Adder's Tongue

Eastern Redcedar

Cedar-Apple Rust Gall

Cedar Glade Animals

A Brief Sampling

Cedar Glade Animals

Though most people's primary interest in cedar glades is as a place to observe wildflowers, a remarkable number of animals call the glades home as well. The surrounding oak-hickory and cedar forests provide habitat for deer, coyote, fox, and other woodland creatures one might expect to encounter. In the grassy zone at the glade edge moles burrow their way through the deeper soils in search of grubs, which incidentally are beetle larvae. Field mice can be found taking refuge in hollow spaces in rotting logs. An assortment of hawks as well as crows, redwing blackbirds and cardinals call loudly from the treetops in all parts of the glades. In late summer look to find goldfinches eating seeds from the various species of cone flowers. Though this section does not offer an exhaustive catalog of cedar glade animals I have included some species that may be unexpected or at least may not be quite so familiar at first site.

Antrostomus carolinensis

Chuck-Will's-Widow - North America's largest member of the Nightjar family the plaintive four-note call of this bird can be heard at dusk and after dark in oak-hickory forests and other dry forest types throughout the Southeast. However their perfect camouflage makes Chuck-Will's-Widow difficult to spot when sitting motionless on the ground or perched on a branch. Photographed May 26, 2012 at Flat Rock Cedar Glades, Rutherford County, TN.

Crotalus horridus

Timber Rattlesnake - These venomous snakes bask quietly at the edge of cedar glades as well as in surrounding forests. They rarely move and will not strike unless provoked or stepped on. In recent years there has been a growing perception that many rattlesnakes no longer make their characteristic rattling sound in advance of a strike. Perhaps snakes with a phenotypic propensity to not rattle have avoided being killed by humans and thus are better able to reproduce. Current research is inconclusive. Rattlesnakes currently face the very real threat of a fungal disease caused by *Crysosporium* spp. which interferes with their ability to molt and may affect hibernation habits. Photographed September 9, 2013, Rocky Branch Road, Wilson County, TN.

Terrapene carolina carolina

Eastern Box Turtle - These land turtles are well documented to live over 100 years in the wild but in the last 30 years their populations have seen over a thirty percent decline, largely due to their propensity to cross busy highways. Coupled with a low birth rate this has caused their conservation status to be listed as Vulnerable by the International Union for Conservation of Nature. It is illegal and usually unsuccessful to take them out of the wild. Collected turtles will usually die within a matter of days. Photographed August 11, 2012 at Flat Rock Cedar Glades, Rutherford County, TN.

Chuck-Will's-Widow

Timber Rattlesnake

Eastern Box Turtle

135

Sceloporus undulatus

Northern Fence Lizard - Lift are large rock on a hot summer day in a cedar glade and you are liable to see this common lizard darting about. They will even run up your leg! With their brown and grey scales Northern Fence Lizards blend in well with rocky ground. However in summer the sides of the belly on males may be greenish-blue. Photographed October 26, 2013 at Stones River National Battlefield, Rutherford County, TN.

Eurycea lucifuga

Cave Salamander - despite their name these salamanders can be found well away from caves, inhabiting rock ledges, moist logs, and under damp leaf litter. This one was found under a large rock at the woodland edge of a limestone glade streamside meadow. Cave salamanders secrete a toxin from their skin when disturbed. Photographed April 12, 2013 at Flat Rock Cedar Glades, Rutherford County, TN.

Vaejovis carolinianus

Southern Unstripped Scorpion or Southern Devil Scorpion - these scorpions have a sting similar to a red wasp. They can be found under rocks or by shining flashlights at night and watching for light to fluoresce off their skin. The female carries the young on her back. When conditions become too dry these scorpions will occasionally invade people's homes. Photographed October 3, 2009 at Flat Rock Cedar Glades, Rutherford County, TN

Northern Fence Lizard

Cave Salamander

Southern Devil Scorpion

137

Pisaurina mira

Nursery Web Spider - Similar to wolf spiders these spiders enclose their egg sac in a silk bag where the young stay until after their first molt -Photographed April 13, 2013 at Flat Rock Cedar Glades Rutherford County, TN

Misumenoides formosipes

White-Banded Crab Spider - this could also be the Goldenrod Crab Spider (*Misumena vatia*). In either case crab spiders lie in wait in flowers where they catch unsuspecting insect prey, usually pollinators, with their front legs - photographed June 3, 2011 at Flat Rock Cedar Glades, Rutherford County, TN

Argiope aurantia

Black and Yellow Garden Spider, Zipper Spider - these large spiders are commonly found in forests around cedar glades and, as their common name implies, around gardens. Though large and terrifying to arachnophobes these spiders will not bite unless handled. Their venom is harmless to humans but may contain polyamines which have been used in cancer treatments though I would not go so far to say their bite is beneficial - photographed September 9, 2013 Rocky Branch Road, Wilson, County, TN

Nursery Web Spider

White-Banded Crab
Spider

Black and Yellow Garden
Spider

Glossary

Achene - an indehiscent fruit, usually flattened. Typical of Asteraceae, ex: sunflower seeds

Actinomorhic - refers to a regular flower; flower parts are radially symmetrical

Anther - the part of a flower where pollen originates; borne atop a filament

Axil (Axillary) - The angular space between the petiole and stem

Basal rosette - lower leaves radially arranged around a central point at ground level

Berry - a wet fruit consisting of a fleshy pericarp usually enclosing multiple seeds, ex: tomato

Bilabiate - a zygomorphic flower in which two lobes of the corolla differ from the others; common in members of the mint family

Bipannately compound - leaves have leaflets arising from a petiole which arises from a pinnate petiole

Carpel - a modified leaf bearing ovules, can be singular in a simple pistil or a one chamber of a compound pistil.

Calyx - a collective term for the sepals

Corolla - a collective term for the petals

Corymb - a flat-topped inflorescence with flowers borne on multiple petioles; outside flowers open first

Cyme - a determinate inflorescence, often flattened, in which the center flowers open first

Determinate - an inflorescence in which the center or top flowers open first with outer or lower flowers opening afterward; growth stops with opening of first flowers

Dicotyledon - "dicot"; a plant in which two leaves emerge from the seed at germination; characterized by netted leaf venation with flower parts in fours or fives

Disjunct - a population of a species occurring outside the normally given range where the species may be common

Dihiscent - describes a fruit that opens at maturity to release seeds

Disk flowers - in Asteraceae, the central portion of the flower head; flowers tubular, clustered and small, contrasting with the more showy ray flowers

Filament - the thin structure upon which the anther sits

Follicle - a fruit originating from one carpel; dehiscent along one suture.

Funnelform - a tubular corolla open at the top and gradually tapering toward the base

Glabrous - smooth, lacking hairs

Glaucous - indicates a smooth, waxy, often whitish coating on leaves

Hirsute - refers to straight hairs on a leaf or stem

Indeterminate - an inflorescence which continues to grow upward or outward as new flowers bloom

Inflorescence - the structure on a plant on which a cluster of flowers are borne

Lanate - covered with course, thick, wool-like hairs

Legume - a dehiscent fruit with a single carpel; ex: a green bean

Monocotyledon - "monocot"; plants in which one leaf emerges from the seed at germination; characterized by parallel leaf venation and flower parts in threes

Nutlet - a small, dry indehiscent fruit with thickened walls

Ovary - the lower portion of the pistil where eggs form; once fertilized it develops into a fruit

Palmately compound - leaflets radiating from a single point

Papilionaceous - literally meaning like a butterfly, this term is used to describe the shape of pea-like flowers often found Fabaceae

Pappus - bristles or hair-like structures on achenes which aid in wind dispersal

Pedicel - a small stem supporting a single flower

Peduncle - a stem supporting an inflorescence

Perianth - a collective term for the corolla and calyx

Pericarp - the part of a fruit surrounding the seeds

Petals - floral leaves around the reproductive organs of a flower; often showy but sometimes greatly reduced or lacking

Petiole - the stalk that attaches a leaf to a stem. A petiolule supports a leaflet.

Pinnately compound - leaflets originating along a longitudinal axis like a feather.

Pistil - the female reproductive organ of a flower including the ovary, style, and stigma; may be simple or compound.

Pollen - structures borne on the anther which produce sperm; often with a highly textured surface specific to the species

Raceme - an unbranched inflorescence with flowers borne on pedicels

Ray flowers - the horizontal, flattened, often colorful flowers on the outer margins of plants in the Asteraceae

Recurved - refers to petals or sepals curving back, away from the axis of the other flower parts

Reflexed - tips pointing below the plane of the flower head.

Salverform - a narrow, tubular corolla opening at the top like the flaps on a box

Scabrous - having a rough or gritty texture like sand paper

Schizocarp - a fruit which splits between carpels forming one seeded portions.

Sessile - lacking a petiole or pedicle; coming directly off stem

Sepals - the outer floral leaves beneath the petals; often green and smaller than the petals but occasionally showy

Silicle - a less elongated silique

Silique - a dehiscent fruit which is often elongated and cylindrical containing many seeds; opens longitudinally along two valves leaving a false partition

Simple - one petiole giving rise to one leaf

Spadix - a club-like spike of inconspicuous flowers; ex: Jack-in-the-Pulpit

Spathe - a leaf-like bract surrounding a peduncle or pedicel; common in monocot

Spikelet - refers to the inflorescence in grasses

Stamen - the male organ of a flower consisting of a filament and an anther

Stigma - the top of the pistil where pollen is caught or deposited

Style - the tube between the stigma and ovary down which sperm travel when released from pollen

Succulent - a stem or leaf which is thick and fleshy to conserve water

Tepal - the lowermost part of the perianth when petals and sepals are not clearly differentiated

Umbel - an inflorescence where each pedicel is attached at the same point; umbrella-like

Zygomorphic - refers to an irregular flower in which flower parts vary, often exhibiting bilateral symmetry

Bibliography

Baskin, J.M. and C.C. Baskin. 1999. Cedar glades of the southeastern United States. Pp. 206-219 in R.C. Anderson, J.S. Fralish, and J.M. Baskin (eds.). *Savannas, barrens, and rock outcrop plant communities of North America.* Cambridge University Press, Cambridge.

Baskin, J.M. and C.C. Baskin. 2004. History of the use of "cedar glades" and other descriptive terms for vegetation on rocky limestone soils in the central basin of Tennessee. *Botanical Review* 70: 403-424

Duke, J.A. 1997. *The Green Pharmacy.* Rodale Press, Emmaus, PA.

Gattinger, A. 1901. *The Flora of Tennessee and a Philosophy of Botany.* Gospel Advocate, Nashville, TN.

Hardin, J.W., D.J. Leopold, F.M. White. 2001. *Harlow and Harrar's Textbook of Dendrology,* Ninth edition. McGraw-Hill, New York.

Hemmerly, T.E. 1990. *Wildflowers of the Central South.* Vanderbilt University Press, Nashville, TN.

Horn, D., T. Cathcart, T.E. Hemmerly, and D. Duhl. 2005. *Wildflowers of Tennessee, the Ohio Valley, and the Southern Appalachians.* Lone Pine Publishing, Auburn WA.

Jones, R.L. 2005. *Plant Life of Kentucky.* University of Kentucky Press, Lexington, KY.

Lewis, W.H. and M.P.F. Elvin-Lewis. 2003. *Medical Botany,* Second edition. John Wiley and Sons, Hoboken , NJ.

Quarterman, E. 1989. Structure and dynamics of the limestone cedar glade communities in Tennessee. *Journal of the Tennessee Academy of Science* 64: 155-158.

Somers, P., L.R. Smith, P.B. Hamel, and E.L. Bridges. 1986. Preliminary analyses of plant communities and seasonal changes in cedar glades of middle Tennessee. *Association of Southeastern Biologists Bulletin* 33: 178–192.

Sutton, A. and M. Sutton. 1997. *Eastern Forests*. Alfred A. Knopf, New York.

Websites
Center for Cedar Glade Studies   http://capone.mtsu.edu/gladectr/
NatureServe   http://www.natureserve.org/

Index

Billy Plant III earned his Master's degree in biology from Middle Tennessee State University. He has been a naval officer, a land surveyor, a biological technician with the National Park Service, and a utility forester. He is the author of *Mother Earth and Other Pretty Girls* and his research is currently being reviewed for publication in *Castanea: Journal of the Southern Appalachian Botanical Society*. He lives in Murfreesboro, Tennessee.